EasyTerms™
Terminology Guidebook
for College-Level Biology

Copyright 2009, Ed Creager

This Biology edition of EasyTerms is the first in a series of simple-to-use, college-level glossary guidebooks. In this handy reference book, terms are arranged alphabetically within appropriate topic areas.

Although they were originally intended for college students, many High School students will also find these guidebooks helpful as they prepare for college.

Other topics to be covered in forthcoming editions:

- Biochemistry
- Botany
- Business Management
- Cell Biology
- Ecology
- Genetics
- Human Anatomy and Physiology
- Microbiology
- Nursing
- Nutrition
- Psychology
- Zoology

EasyTerms can help support your educational advancement and can boost the vocabulary of almost anyone who reads it.

For more information on these and other publications, please visit the author's site:

www.edcreager.blogspot.com

... and please note the author's "signature book" entitled,

"The Money-Saving Idea Book: Inside Tips for Starving Students, Frugal Seniors and Every Financial Survivor."

("The Money-Saving Idea Book" © and TM, Ed Creager, 2009.)

Foreword

This edition of EasyTerms is one in a series of simple-to-use, college-level* glossary guidebooks. In this handy reference book, terms are arranged alphabetically within appropriate topic areas. The complete index makes it easy to find any term and its definition.

* These books can also help High School students prepare so that, before they attend college, they'll already know a considerable amount of the terminology they'll need.

A substantial number of the terms defined here have additional definitions outside the scope of the subject being covered. More general definitions and additional meanings, if sought, are to be found in less specialized publications such as dictionaries and encyclopedias.

Please check the website of the author...

www.edcreager.blogspot.com

... for more information on other available books.

Other subjects to be covered in forthcoming editions of EasyTerms include, but are not limited to:

- Biochemisty

- Cell Biology

- Human Anatomy & Physiology

- Nursing

- Nutrition

Important Notice:

The resources provided hereby, including websites, books and related materials, are intended to provide accurate information regarding the subject matter. All products and services are provided with the understanding that neither the author nor the publisher is engaged in rendering legal, accounting, or other professional advice. If expert assistance is needed, the services of a competent professional should be obtained.

EasyTerms™
Terminology Guidebook
Table of Contents

The terms that follow are divided into the topics shown below. The page number on which the topic begins is given. Within each topic, the terms are arranged alphabetically.

Introduction	1
Introduction to Chemistry	6
Chemistry of Cells	10
Cells	13
Cellular Respiration	20
Photosynthesis	22
Gas Exchange	24
Nutrition and Digestion	28
Transport	38
Body Defenses	49
Excretion and Fluid Regulation	54
Neurons and Sensory Systems	57
Support and Movement	73
Hormones	82
Animal Behavior	86
Animal Reproduction	89
Classical Mendelian Genetics	97
Molecular & Population Genetics	101
Development	106
Evolution	114
Viruses and Prokaryotes	118
Protists and Fungi	120
Plants and Plant Reproduction	123
Animals	130
Ecology and Environment	136
Index	145

Introduction

1. abdomen

Body region between the diaphragm and the pelvis.

2. abdominal

Of the abdomen.

3. anatomy

The study of structure.

4. anterior

Ventral, toward the belly.

5. biology

The study of life.

6. biosphere

An interconnected system over the earth's surface in which organisms exist.

7. botany

The study of plants.

8. community

A set of interacting organisms living in a given location.

9. complementarity

A relationship in which structures and their functions are mutually reinforcing.

10. continuity of life

The transfer of life from parent to offspring.

11. cranial

Pertaining to the cranium.

12. data

Observations from an experiment.

13. deductive reasoning

Reasoning from a general statement to a specific case.

14. development

Process of increasing in complexity.

15. distal

Most distant from the point of origin of a structure.

16. diversity

The quality of variation among a class, such as living organisms.

17. dorsal

Toward the back; posterior in humans.

18. ecosystem

All the organisms in a natural setting and their physical environment.

19. environment

Changeable surroundings around a living organism.

20. excitability

Responsiveness to a stimulus.

21. external

On the outside.

22. feedback

The use of the output of a process to influence the process.

23. growth

Increase in size.

24. homeostasis

The maintenance of a narrow, tolerable range of internal conditions.

25. homeostatic system

A control system that helps to maintain homeostasis.

26. hypothesis

A possible answer to a question; a possible explanation for observations that can be used to predict future outcomes.

27. inductive reasoning

Development of a general statement from a collection of observations.

28. internal

Inside or within.

29. internal environment

The environment around cells, but within the body.

30. interstitial

Concerning spaces between cells.

31. kingdom

A major taxonomic subdivision of living organisms.

32. lateral

On or toward the side.

33. liter

The basic metric unit of fluid volume; approximately 1.06 quarts.

34. macromolecule

A very large molecule such as a protein or nucleic acid.

35. mass

The amount of matter in an object.

36. median

Toward or in the middle.

37. meter

The basic metric unit of length; approximately 39.37 inches.

38. microscope

An instrument for observing structures too small to see with the naked eye.

39. natural selection

A process by which well adapted organisms survive and reproduce in larger numbers than less well adapted ones.

40. organ

A structure made up of several tissues that carries out particular functions; component of a system.

41. organism

A living thing.

42. organizational complexity

A concept that concerns the structural levels of an organism.

43. peripheral

Outer or away from the center.

44. physiology

The study of life functions.

45. posterior

Toward the rear, dorsal in humans.

46. proximal

Nearest the point of origin of a structure.

47. reproduction

Process by which offspring arise.

48. responsiveness

Ability to react to a stimulus.

49. sagittal

A plane dividing right and left sides.

50. scientific method

A method of collecting and testing data pertaining to a scientific question.

51. theory

An explanation that accounts for many observations.

52. thorax

Chest; body region above diaphragm.

53. tissue

A group of similar cells, including intercellular substances, that carry out a particular function.

54. transverse

Crosswise.

55. ventral

Of or toward the belly.

56. viscera

Internal organs.

57. weight

The amount of gravitational force between two objects.

58. zoology

The study of animals.

59. acid

An ionizing substance that donates hydrogen ions.

60. alkaline

Basic, able to accept hydrogen ions.

61. anion

A negatively charge ion.

62. atom

Smallest particle that retains properties of an element.

63. atomic number

The number of protons in the nucleus of an atom.

64. atomic weight

The total number of protons and neutrons in an atom; the average number if there are isotopes of the element.

65. base

An ionizing substance that accepts hydrogen ions or reacts with an acid to form a salt.

66. buffer

A substance that resists pH change by holding or releasing hydrogen ions in a solution.

67. catalyst

A substance that increases a chemical reaction rate.

68. cation

A positively charged ion.

69. colloid

Glue-like; a particle in a colloidal dispersion.

70. colloidal dispersion

A state of matter with small particles suspended in a medium.

71. compound

A substance with two or more elements combined in definite proportion.

72. covalent bond

A chemical bond formed by shared electrons between two atoms.

73. dehydration

Removal of water.

74. denaturation

An alteration in the shape and properties of a protein molecule.

75. electron

A negatively charged particle that continually moves around the nucleus of an atom.

76. element

A fundamental unit of matter.

77. endergonic

Requiring energy, as in a chemical reaction.

78. entropy

Tendency toward chaos or disorder.

79. exergonic

Releasing energy, as in a chemical reaction.

80. gram molecular weight

The quantity of a substance (in grams) equal to its molecular weight.

81. hydrogen bond

Weak covalent bond between hydrogen and another element, such as oxygen or nitrogen.

82. hydrolysis

The splitting of a molecule with the addition of water.

83. hydrophilic

Attacted to water.

84. hydrophobic

Tending to avoid water.

85. ion

A charged atom or group of atoms.

86. ionic bond

A chemical bond with atoms held together by the attraction of unlike charges.

87. isomer

A molecule having the same kinds and number of atoms as another molecule, but arranged differently.

88. isotope

An atom having a different number of neutrons than certain other atoms of the same element.

89. kinetic

Energy of motion.

90. mixture

Two or more substances combined in any proportions and retaining their individual properties.

91. mole

A gram molecular weight.

92. molecule

The smallest quantity of a substance that retains its chemical properties.

93. neutron

An uncharged particle in the nucleus of an atom.

94. nonpolar

Lacking charged regions.

95. nucleus

Central part of an atom or a cell.

96. organic

Containing carbon.

97. oxidation

Addition of oxygen or loss of electrons in a chemical reaction.

98. pH

The negative logarithm of the hydrogen ion concentration; a scale for expressing acidity or alkalinity.

99. polar compound

A molecule having a charged area or polarity.

100. potential energy

Energy due to position and capable of being released, as in a rock at the top of a hill.

101. proton

A positively charged particle in the nucleus of an atom.

102. radiation

Spreading from a center; giving off electromagnetic particles and waves.

103. reactant

A substance that enters into a chemical reaction.

104. reduction

Gain of an electron or loss of oxygen in a chemical reaction.

105. secretion

A cell product; the active transport of substances from the blood to the kidney filtrate.

106. specific heat

The amount of heat needed to increase the temperature of a specific volume of substance one degree Celsius.

107. trace element

A chemical element normally present in very small amounts in the body.

108. valence

An ion's charge.

Chemistry of Cells

109. adenosine triphosphate

An important energy storage molecule.

110. amino acid

A molecule having both acid and amino functional groups.

111. carbohydrate

An organic compound having several alcohol groups and an aldehyde or ketone group.

112. deoxyribonuclease

An enzyme that digests DNA.

113. deoxyribonucleic acid (DNA)

A nucleic acid in chromosomes that directs protein synthesis and transmits genetic information to a new generation.

114. disaccharide

A molecule having two sugar (saccharide) units held together by a glycosidic bond.

115. enzyme

A protein that increases the rate of a chemical reaction in a living organism.

116. fatty acid

A long hydrocarbon chain with a carboxyl group at one end.

117. functional group

A component of a molecule that participates in a chemical reaction.

118. glycine

An amino acid with the simplest chemical structure.

119. glycolipid

A molecule that contains both carbohydrate and lipid components.

120. glycoprotein

A molecule that contains both carbohydrate and protein components.

121. lecithin

A phospholipid characteristic of animal tissues.

122. lipid

Fat or fatlike substance.

123. monosaccharide

A simple sugar.

124. nucleic acid

A polymer of nucleotides; DNA or RNA.

125. nucleotide

A molecule having a nitrogenous base, a 5-carbon sugar, and one or more phosphates.

126. peptide bond

A chemical bond between the amino group of one amino acid and the carboxyl group of another.

127. phospholipid

A lipid made of glycerol, fatty acids, and phosphoric acid.

128. polymer

A molecule consisting of repeating units.

129. polypeptide

A chain of amino acids held together by peptide bonds.

130. polysaccharide

A molecule consisting of many saccharide units connected by glycosidic bonds.

131. protein

A polymer of amino acids.

132. ribonucleic acid (RNA)

A nucleic acid made from information in DNA that is involved in protein synthesis.

133. saturated fatty acid

A fatty acid lacking double bonds in the carbon chain and being saturated with hydrogen.

134. saturation

Condition of having all chemical affinities satisfied.

135. specificity

The attribute of being specific.

136. stereoisomer

Compound having the same kind and number of atoms as another compound, but in a different spatial arrangement.

137. steroid

A lipid with a complex four-ring structure.

138. triglyceride

A triacylglycerol (glycerol and three fatty acids).

139. unsaturated fatty acid

Fatty acid with pairs of hydrogen atoms replaced by double bonds in the carbon chain.

140. uridine triphosphate (UTP)

A high energy molecule.

Cells

141. active transport

Transport of a substance against a gradient using a carrier molecule, enzyme, and cellular energy.

142. adsorptive endocytosis

Entry of a substance into a cell after attaching to the cell membrane.

143. anaphase

A mitotic stage during which chromosomes move apart.

144. aster

Short microtubules at the ends of a spindle in a dividing animal cell.

145. binding site

A site where a particular molecule binds to a membrane or other structure.

146. cell

A basic functional unit of a living organism.

147. cell cycle

A repetitive sequence of events involving DNA replication and cell division.

148. cell membrane

Lipid and protein compounds that form the boundary of a cell.

149. cell theory

A theory stating that living things are composed of cells.

150. centriole

One of a pair of intracellular bodies that participate in forming a mitotic spindle.

151. chloroplast

A membrane-bound organelle containing chlorophyll and enzymes needed for photosynthesis.

152. chromatin

Nuclear material that condenses into distinct chromosomes during cell division.

153. chromosome

In a human cell, one of 46 nuclear structures made of DNA and protein.

154. cilium

A tiny hairlike projection found on some epithelial cells.

155. coenzyme

A substance that works with an enzyme in activating chemical reactions.

156. cytokinesis

Division of the cytoplasm that follows division of a nucleus.

157. cytoplasm

Cell substance, excluding the nucleus.

158. cytoplasmic streaming

Movement of chloroplasts and other structures within a cell's cytoplasm.

159. cytoskeleton

The organelles forming a cell's internal framework.

160. cytosol

The fluid part of cytoplasm that suspends organelles.

161. DNA polymerase

An enzyme that increases chain length in DNA synthesis.

162. DNA replication

Synthesis of new DNA according to information in an existing DNA template.

163. endoplasmic reticulum

A membranous vesicular network within a cell.

164. eukaryotic

Having a nucleus and membrane-bound organelles.

165. extracellular

Outside a cell.

166. facilitated diffusion

Diffusion down a gradient on carrier molecule but not requiring cellular energy.

167. filtration

Passage of a fluid across a membrane by mechanical pressure.

168. flagellum

A movable hairlike process on a cell.

169. fluid-mosaic model

A model of molecular arrangements in a cell membrane.

170. gel

A liquid state in a colloidal dispersion.

171. genetic code

The three-base sequences in messenger RNA derived from a DNA template that determine amino acid order in proteins.

172. Golgi apparatus

Membranous vesicles clustered in cells that complete synthesis of secretions.

173. gradient

The rate of change in the magnitude of concentration, pressure, or other variable.

174. haploid

Having one of a pair of chromosomes.

175. hydrostatic pressure

Force exerted by a fluid.

176. hyperosmotic

Having higher osmotic pressure than a reference solution.

177. hypertonic

Causing movement of water out of cells.

178. hyposmotic

Having lower osmotic pressure than a reference solution.

179. **hypotonic**

Causing movement of water into cells.

180. **interphase**

A cell cycle stage during which the cell is not dividing.

181. **intracellular**

Within a cell.

182. **intrinsic**

Entirely within.

183. **isosmotic**

Having the same osmotic pressure as a reference solution.

184. **isotonic**

Causing no net water movement across a cell membrane.

185. **karyotype**

Arrangement of chromosomes from a cell in pairs and in a fixed order.

186. **ligand**

That which binds to a receptor.

187. **lysosome**

Membrane-bound organelle that contains digestive enzymes.

188. **malignancy**

A tendency to become more virulent; a cancerous growth.

189. **messenger RNA**

A nucleic acid that carries information in the form of codons for the synthesis of a protein.

190. **metaphase**

A mitotic stage during which chromosomes align along the equator of a cell.

191. **metastasis**

The transfer of disease from one organ to another.

192. microfilament

A small, hollow protein fiber in cytoplasm that aids in movement or forms part of a cytoskeleton.

193. microtubule

A cylindrical organelle that forms part of a cell's mitotic spindle.

194. mitochondrion

An organelle that contains enzymes for oxidative and energy-capturing processes.

195. mitosis

Nuclear division that produces two identical nuclei.

196. nuclear

Of the nucleus.

197. nucleolus

A body containing RNA within a nucleus.

198. nucleoplasm

The substance of a nucleus.

199. organelle

A tiny function unit within a cell.

200. osmolarity

A solution's osmotic concentration determined by the number of osmotically active particles it contains.

201. osmosis

Diffusion of water through a membrane from its own higher to a lower concentration.

202. osmotic pressure

Pressure created by osmosis.

203. passive transport

A process that moves substances without the energy expenditure by the organism.

204. peroxisome

An organelle containing oxidative enzymes.

205. plasma membrane

Membrane forming the boundary of a cell.

206. plasmolysis

The process by which a plant cell protoplast shrinks from its cell wall due to water loss in a hypertonic solution.

207. prokaryotic

Lacking a nucleus and membrane-bound organelles.

208. prophase

The first mitotic stage during which the chromosomes become distinct.

209. protoplasm

Cell substance; literally, first formed.

210. receptor

A specific site with which a specific substance can bind; cell that responds to signals sensed from the environment.

211. remission

The abatement of disease symptoms or the period during which it occurs.

212. ribosome

An organelle containing ribonucleic acid and protein where protein synthesis occurs.

213. selectively permeable

A membrane property that allows passage of some substances while preventing passage of others.

214. sex chromosome

A chromosome associated with maleness or femaleness; X or Y chromosome in mammals.

215. sodium-potassium pump

Mechanism that actively moves Na ions out of cells and K ions into them against gradients.

216. sol

A liquid state of a colloidal dispersion.

217. solute

A dissolved substance.

218. **solution**

A liquid containing dissolved substances.

219. **solvent**

A substance in which other substances can dissolve.

220. **spindle fiber**

Microtubules in eukaryotic cells involved in the movement of chromosomes during mitosis and meiosis.

221. **surface tension**

Resistance to rupture by the surface film of a liquid.

222. **surface-to-volume ratio**

The surface area of a structure divided by its volume.

223. **telophase**

The last mitotic stage during which nuclei reform.

224. **teratogen**

An agent that causes defective embryonic development.

225. **tonicity**

The degree to which fluid can move into or out of cells.

226. **tubulin**

A protein that forms intracellular microtubules.

227. **tumor necrosis factor**

A substance that causes degeneration and death of tumor cells.

228. **turgor pressure**

Pressure created in plant cells by uptake of water from a hypotonic solution.

Cellular Respiration

229. aerobic

In the presence of oxygen.

230. amylase

An enzyme that digests starch.

231. anabolic

Of anabolism.

232. anabolism

Synthetic, energy using process.

233. anaerobic

Lacking oxygen.

234. catabolism

Breakdown of molecules that makes energy available.

235. deamination

Removal of an amino group.

236. electron transport system

Enzymes and coenzymes in cristae of mitochondria that move electrons from substrates to oxygen.

237. fermentation

An anaerobic metabolic process in which carbohydrate is broken down to alcohol and other simple molecules.

238. flavin adenine dinucleotide (FAD)

A coenzyme that carries hydrogen.

239. gluconeogenesis

Metabolic pathway that makes glucose from noncarbohydrate substances.

240. glycogenesis

Metabolic pathway for glycogen synthesis.

241. glycogenolysis

Metabolic pathway for glycogen breakdown.

242. glycolysis

Metabolic pathway for breakdown of glucose to pyruvic acid.

243. Krebs cycle

Metabolic pathway that oxidizes acetyl-CoA; citric acid cycle; tricarboxylic acid cycle.

244. lipoprotein

A molecule made of lipid and protein.

245. metabolism

All chemical reactions in a living organism.

246. nicotinamide adenine dinucleotide (NAD)

A coenzyme that transports hydrogen atoms or electrons in oxidation-reduction reactions.

247. oxidative phosphorylation

Capture of energy in ATP during oxidative metabolism.

248. phosphorylation

Binding of a phosphate group to a molecule.

249. transamination

Transfer of an amino group from one molecule to another.

250. turnover

Reuse of a substance made available by a catabolic reaction.

Photosynthesis

251. anabolism

Synthetic, energy using process.

252. catabolism

Breakdown of molecules that makes energy available.

253. chemosynthesis

A proces by which large molecules are made from small ones using energy from other chemicals.

254. chlorophyll

A green pigment capable of capturing light energy.

255. cyclic photophosphorylation

The capture of energy in chloroplasts with the formation of ATP through the activity of the cell's cytochrome system.

256. dark reaction

A part of photosynthesis that can occur in either light or dark and that transfers energy from light reactions.

257. granum

A stack of thylakoids within a chloroplast.

258. guard cell

One of a pair of cells that regulate openings in leaf epidermis.

259. light reaction

Events in photosynthesis that capture energy and occur in light.

260. metabolism

All chemical reactions in a living organism.

261. noncyclic photophosphorylation

The capture of energy in chloroplasts with the formation of ATP and reduced NADP.

262. photophosphorylation

The addition of phosphate groups and high energy bonds to a molecule during the capture of light energy.

263. photosynthesis

The process by which organisms capture light energy from the environment and store it in a usable form.

264. spongy parenchyma

Photosynthetic tissue, which has loosely arranged cells.

265. stoma

A pore in lower leaf epidermis through which gases diffuse into and out of mesophyll spaces.

266. thylakoid

Parallel flattened sacs that form part of the membrane structure of a chloroplast.

Gas Exchange

267. Adam's apple

Thyroid cartilage of larynx, which is prominent in males.

268. alveolus

One of many small air sacs in the lungs and in secretory parts of some glands; a tooth socket.

269. apical

Located at the apex or tip of a structure.

270. asthma

A disorder in which constriction of bronchioles causes difficulty in breathing.

271. book gill

A respiratory organ found in certain crabs.

272. book lung

A respiratory organ found in spiders.

273. Boyle's law

Pressure exerted by a gas is inversely proportional to its volume.

274. bronchiole

One of many small tubes in the lungs.

275. bronchitis

Inflammation of bronchi.

276. bronchus

One of many tubes leading from the trachea to the lung.

277. cardiopulmonary resuscitation (CPR)

A method for maintaining blood flow and gas exchange in a person with no heart beat and no breathing.

278. Dalton's law

Each gas in a mixture exerts a partial pressure that is independent of other gases.

279. emphysema

A disorder characterized by destruction or dilation of walls of alveoli.

280. Eustachian tube

A passage that connects the middle ear and the pharynx.

281. expiration

Exhaling; breathing out.

282. gas exchange

The diffusion of gases across membranes as when oxygen enters and carbon dioxide leaves blood.

283. gill

An aquatic animal's respiratory organ.

284. glottis

A slitlike opening from the pharynx to the larynx.

285. Henry's law

A gas dissolves in a liquid in proportion to its solubility and its partial pressure.

286. inspiration

Breathing in.

287. larynx

Voice box.

288. lenticel

A group of loosely arranged cork cells that allow gas exchange.

289. mantle

A membranous organ that secretes the shell of a mollusk.

290. mantle cavity

A space between the body and the mantle of a mollusk; contains the gills.

291. operculum

A protective covering over the gill chamber of a fish.

292. palate

Flat plate-like roof of the mouth.

293. partial pressure

Pressure exerted by one gas in a mixture of gases.

294. pharynx

Throat.

295. pressure

Force due to compression.

296. pulmonary

Of the lungs or the blood vessels carrying blood to and from gas exchange membranes.

297. respiration

The processes of ventilation (breathing) and gas exchange.

298. respiratory center

A neural center in the brain stem that regulates respiration.

299. respiratory pigment

A cytochrome or other pigment that increases the oxygen carrying capacity of blood or another tissue.

300. spiracle

A respiratory passageway in an arthropod.

301. spirometry

Measurement of gas volumes entering or leaving the lungs.

302. surfactant

A phospholipid that reduces surface tension.

303. swim bladder

A dorsal sac in the body cavity of a fish that allows it to regulate its bouyancy.

304. tetrapod

Having four appendages.

305. tonsil

An aggregate of pharyngeal lymphatic tissue.

306. trachea

Passage from the larynx to the bronchi.

307. ventilation

Movement of gases between the lungs and the environment.

308. vital capacity

The largest gas volume that can be expired after a maximal inspiration.

309. vocal

Of the voice.

Nutrition and Digestion

310. **absorption**

Movement of substances across a membrane.

311. **absorptive**

Concerning absorption.

312. **adventitia**

Outermost connective tissue layer on an organ or blood vessel.

313. **alcoholism**

Disease of excessive alcohol intake, which almost invariably leads to increasingly severe problems in daily living.

314. **anorexia nervosa**

A serious neurological disorder in which a person loses weight and becomes emaciated.

315. **anus**

An opening through which wastes exit the digestive tract.

316. **appendicitis**

Inflammation of the appendix.

317. **argentaffin**

Cells in the stomach lining that secrete histamine and serotonin.

318. **ascorbic acid**

Vitamin C.

319. **autotroph**

An organism that makes organic compounds from inorganic substances in the environment.

320. **basal metabolic rate (BMR)**

Amount of energy used to maintain life in an awake, resting individual.

321. **basal metabolism**

The process of using energy from nutrients to maintain life in an awake, resting state.

322. **bicuspid**

Having two points or cusps.

323. bile

Liver secretion that aids in digestion by emulsifying fats.

324. bolus

A mass.

325. bulimia

Binge eating, usually followed by self-induced vomiting.

326. calorie

Quantity of heat to raise the temperature of one gram of water one degree Celsius.

327. carnivore

An animal whose diet consists mainly of other animals.

328. carotene

Yellow substance that usually has vitamin A activity.

329. CCK-PZ (cholecystokinin-pancreozymin)

An enteric hormone that stimulates the gallbladder to release bile and the pancreas to secrete enzymes.

330. cecum

Blind pouch.

331. cellulose

A polysaccharide found in the structure of many plants.

332. cementum

Material surrounding dentin in the root of a tooth.

333. chemolithotroph

Organisms that derive energy from inorganic compounds.

334. chemotroph

An organism that gets energy from oxidizing inorganic or organic matter.

335. chylomicron

A particle consisting of lipids and proteins made in the intestinal mucosa and released into lacteals.

336. chyme

Semiliquid, partially digested food leaving the stomach.

337. chymotrypsin

A proteolytic enzyme from the pancreas.

338. colon

Large intestine from the cecum to the rectum.

339. core body temperature

Temperature deep within the body.

340. cuspid

A point or tapering projection.

341. cystic fibrosis

An inherited disorder in which thick mucus blocks respiratory and pancreatic passageways.

342. deciduous teeth

Nonpermanent teeth; 'baby teeth'.

343. defecation

Passage of wastes from the rectum outside the body.

344. deglutition

Swallowing.

345. dentin

Bonelike substance of a tooth located beneath the surface.

346. diabetes mellitus

A disorder due to lack or inactivity of insulin that allows glucose to accumulate in the blood and urine.

347. diarrhea

Excessively frequent, fluid bowel movements.

348. digestion

Breakdown of large molecules into smaller ones.

349. duodenum

A short part of the small intestine adjacent to stomach that receives secretions from the liver and pancreas.

350. emulsification

Process by which bile salts cause fat droplets from foods to break into smaller particles.

351. enamel

Hard covering of a tooth seen above the gumline.

352. epiglottis

Elastic cartilage that closes the glottis.

353. esophagus

Muscular tube between the pharynx and stomach.

354. essential amino acid

Amino acid required in the diet because the body cannot make it.

355. essential fatty acid

Fatty acid required in the diet because the body cannot make it.

356. evaporation

Changing of a substance from liquid to gaseous form.

357. feces

Digestive waste expelled from the rectum through the anus.

358. folacin

Vitamin needed to help transfer single carbon groups.

359. fundus

Part of an organ farthest from its outlet.

360. gallbladder

A sac on the underside of the liver where bile is stored.

361. gastric

Of the stomach.

362. gingiva

Gums.

363. gluten

A protein containing gliadin found in wheat and some other grains.

364. herbivore

An animal whose diet consists mainly of plants.

365. heterotroph

An organism that metabolizes ready-made organic matter.

366. high density lipoprotein (HDL)

A blood particle containg protein and lipid that tends not to deposit cholesterol in blood vessels.

367. hyperthermia

An abnormally high body temperature.

368. hypothermia

An abnormally low body temperature.

369. ileum

Lower part of the small intestine.

370. incisor

A cutting tooth.

371. ingestion

Intake of food or fluid.

372. insectivorous

Pertaining to a plant that increases its nitrogen intake by capturing and ingesting insects.

373. intrinsic factor

Substance secreted by the gastric mucosa required for the transport and absorption of vitamin B12.

374. jejunum

Middle part of the small intestine.

375. kilocalorie

Heat required to raise the temperature of one kilogram of water one degree Celsius.

376. Kupffer's cell

A phagocytic cell in the walls of liver sinusoids.

377. lactase

An enzyme that digests lactose.

378. lacteal

Lymph vessel in a villus of the small intestine.

379. lipase

An enzyme that breaks down lipids.

380. low density lipoprotein (LDL)

A blood particle containing protein and lipic implicated in the deposition of cholesterol in blood vessel walls.

381. macronutrient

Nutrient needed in relatively large amounts.

382. malnutrition

Ill health caused by an inadequate diet.

383. maltase

Enzyme that digests maltase, a disaccharide derived from starch.

384. mesentery

A two-layered membrane continuous with the peritoneum that suspends an abdominal organ.

385. metabolic rate

Rate at which nutrients are oxidized.

386. micelle

A small fat droplet in chyme.

387. micronutrient

A nutrient needed in relatively small quantities.

388. microvillus

A cytoplasmic projection of surface membrane of intestinal epithelial cells.

389. mineral

Inorganic substance.

390. molar

A large grinding tooth; pertaining to the concentration of a solution.

391. motility

Ability to move.

392. mucosa

Mucous membrane lining cavities and passageways.

393. net protein utilization

Proportion of protein eaten that is actually used by cells.

394. niacin

B vitamin used to synthesize the coenzyme NAD.

395. nutrition

The act of providing substances needed for good health through food ingestion.

396. obesity

The condition of having body fat in an amount that is well above the normal amount.

397. omnivore

An animal that eats both plant and animal tissue.

398. pancreas

A digestive gland that secretes enzymes and hormones.

399. pantothenic acid

B vitamin used to synthesize coenzyme A.

400. pepsin

An enzyme that starts breakdown of protein in the stomach.

401. peristalsis

Wavelike, propelling contractions along tubular passageways.

402. peritoneum

A membrane that covers abdominal organs and lines the abdominal cavity.

403. Peyer's patch

An elevated lymphoid tissue mass in the mucosa of the small intestine.

404. photoauxotroph

An organism that uses light energy to manufacture its own food.

405. photolithotroph

An organism that uses light energy to synthesize food from inorganic substances.

406. pulp cavity

A chamber within a tooth that contains blood vessels and nerves.

407. pylorus

Stomach region attached to the small intestine.

408. pyrogen

A substance that causes fever.

409. rectum

Terminal portion of the digestive tract between the colon and the anal canal.

410. riboflavin

Heat-labile B vitamin used to synthesize the coenzyme FAD.

411. ribonuclease

Enzyme that digests RNA.

412. ruminant

A mammal that chews cud.

413. salivary

Of saliva or glands that produce it.

414. scurvy

A disease due to a vitamin C deficiency.

415. secretin

A hormone from the intestinal mucosa that stimulates secretion of bile and pancreatic fluid.

416. sphincter

A ringlike muscle by which a natural orifice opens and closes.

417. stomodeum

Ectodermal evagination from which the mouth and adjacent pharynx form.

418. submucosa

A layer beneath the intestinal mucosa.

419. sucrase

An enzyme that digests sucrose.

420. thermogenesis

Heat generation.

421. thiamine

A water-soluble B vitamin used to synthesize cocarboxylase.

422. tocopherol

A substance with vitamin E activity.

423. total parental nutrition (TPN)

Process of giving all required nutrients by a route other than the digestive tract.

424. trypsin

A proteolytic enzyme released from the pancreas.

425. very low density lipoprotein (VLDL)

A particle in the blood containing much lipid and little protein.

426. villus

Vascular tuft.

427. vitamin A

Vitamin needed to synthesize visual pigments and maintain epithelial cells.

428. vitamin D

A vitamin that facilitates calcium absorption.

429. vitamin E

Vitamin that acts as an antioxidant.

430. vitamin K

Vitamin needed for synthesis of some blood clotting factors.

Transport

431. adhesion

Tendency of unlike molecules to stick together.

432. agranular leukocyte

A white blood cell lacking cytoplasmic granules.

433. albumin

A small protein made in the liver and released into blood.

434. anemia

A hemoglobin deficiency associated with too few erythrocytes or poorly functioning ones.

435. aneurysm

A saclike dilation in an arterial wall.

436. anticoagulant

A substances the prevents blood clotting.

437. aorta

A large artery that carries blood from the left ventricle to other arteries of the systemic circulation.

438. aplastic

Having no tendency to undergo cell division.

439. arteriole

A blood vessel between an artery and capillaries.

440. artery

A large blood vessel carrying blood away from the heart.

441. artificial pacemaker

A device that automatically stimulates the heart and maintains a regular heart rate.

442. atrium

A chamber or entrance.

443. auricle

An earlike appendage.

444. basophil

A leukocyte having granular cytoplasm and able to be stained with a basic dye.

445. blood

Fluid pumped by the heart through a closed system of vessels.

446. Bohr effect

Tendency of high oxygen concentration in the lungs to cause hemoglobin to release carbon dioxide.

447. capillary

A small blood vessel interposed between an arteriole and a venule.

448. carbaminohemoglobin

Hemoglobin to which carbon dioxide is bound.

449. cardiac

Of the heart.

450. cardiovascular

Of the heart and blood vessels.

451. Casparian strip

A waxy material on endodermal plant cells that prevents molecules from leaving the cells.

452. celiac

Related to the abdomen.

453. circulation

The continuous passage of blood through blood vessels from one region of the body to another.

454. closed circulatory system

A continuous system of blood vessels that contain an animal's blood.

455. cohesion

The attraction of water molecules for each other.

456. conduction system

Fibers in heart muscle in which signals coordinate atrial and ventricular contractions.

457. conus arteriosus

A large artery in frogs through which blood leaves the heart.

458. cork cambium

Meristem that produces cork in a plant stem.

459. cross-matching

Comparison of donor and prospective recipient bloods to detect possibilities of agglutination.

460. diapedesis

Squeezing of leukocytes between the cells of capillary walls.

461. dorsal root ganglion

A ganglion containing cell bodies of neurons that carry sensory signals.

462. edema

Excess fluid accumulation in the tissues.

463. electrocardiogram (ECG)

A record of electrical changes detected on the body surface associated with heart contractions.

464. electroencephalogram (EEG)

A record of electrical changes detected on the scalp and associated with brain activity.

465. endocardium

The heart's epithelial lining.

466. endodermis

A single layer of cells that regulates the entry of materials into roots.

467. eosinophil

A granular leukocyte capable of being stained with the dye eosin.

468. erythrocyte

Red blood cell.

469. erythropoiesis

Process of red blood cell formation.

470. erythropoietin

A substance from the kidney that stimulates erythropoiesis.

471. ferritin

A molecule made up of the protein apoferritin and iron.

472. fibrin

A fibrous protein that forms a network in a blood clot.

473. fibrinogen

Inactive fibrin.

474. germinal epithelium

Epithelial cells that divide to form gamete-producing cells.

475. globin

A globular protein found in hemoglobin and certain other biological molecules.

476. globulin

A globular shaped protein, many of which are found in plasma.

477. granular leukocyte

A white blood cells with granular cytoplasm.

478. guttation

The forcing of water from a plant's leaf tips.

479. heart failure

Loss of ability to pump sufficient blood to supply body tissues with nutrients and remove wastes.

480. heart murmur

An abnormal sound caused by turbulence around a defective valve.

481. hematocrit

The proportion of erythrocytes in a volume of blood.

482. hematopoiesis

The formation of blood cells.

483. heme

An iron-containing pigment in hemoglobin that binds oxygen.

484. hemodialysis

The removal of substances from the blood by dialysis.

485. hemoglobin

The oxygen-carrying protein in erythrocytes.

486. hemolysis

Breakdown of erythrocytes with hemoglobin release.

487. hemophilia

An inherited inability to produce a blood clotting factor.

488. hemopoiesis

The formation of blood cells.

489. hemorrhage

Loss of a significant volume of blood.

490. hirudin

An anticoagulant secreted by leeches.

491. hypertension

Excessively high blood pressure.

492. hypotension

Excessively low blood pressure.

493. ischemia

Reduction in blood flow to an area.

494. isovolumetric

Having the same volume.

495. jaundice

Yellowish tone to skin and membranes caused by excess bile pigments in the blood.

496. leaf vein

A vascular structure within a leaf.

497. leukocyte

A white blood cell.

498. lymph

Interstitial fluid in a lymphatic vessel.

499. lymph node

A lymphatic tissue aggregation interposed at intervals along a lymphatic vessel.

500. lymphatic

Concerning lymph or a vessel that carries it.

501. lymphocyte

A leukocyte that participates in an immune response.

502. lymphoid

Resembling lymph or lymphatic tissue.

503. mesophyll

Photosynthetic tissue between the upper and lower surfaces of a leaf.

504. monocyte

A large, phagocytic, agranular leukocyte.

505. myocardium

The thick muscular layer of the heart.

506. neutrophil

A granular leukocyte that fails to stain with either acidic or basic stains.

507. oncotic pressure

Osmotic pressure created by the proteins and other molecules in a fluid.

508. open circulatory system

A system of blood vessels that carries blood between the heart and body cavities.

509. **organ of Corti**

The inner ear structure where sound receptors are located.

510. **oxyhemoglobin**

Hemoglobin to which oxygen is bound.

511. **pacemaker**

An aggregation of cells that spontaneously excite other cells, as in the sinoatrial node.

512. **pericardium**

A sac around the heart.

513. **pericycle**

The outermost tissue of a vascular cylinder from which branch roots and cork cambium arise.

514. **pernicious anemia**

Anemia due to a lack of intrinsic factor and therefore vitamin B12.

515. **phloem**

The carbohydrate transporting tissue of a plant.

516. **plaque**

A sheetlike deposit.

517. **plasma**

The fluid part of blood including inactive clotting factors.

518. **plasmin**

An enzyme built into blood clots as they form that gradually dissolves them.

519. **plasminogen**

Inactive plasmin.

520. **platelet**

A megakaryocyte fragment in blood that participates in blood clotting reactions.

521. **prothrombin**

Inactive thrombin.

522. pulse

Rhythmic expansion and contraction of an artery caused by the pumping action of the heart.

523. Purkinje fiber

The terminal ends of fibers in the heart's conduction system.

524. resistance

Opposition to flow, as in blood vessels.

525. root cap

Protective tissue on the tip of a root.

526. SA node

Part of the heart's conduction system that normally initiates contractions.

527. serum

The fluid part of blood after removal of formed elements and clotting factors.

528. sickle cell anemia

An inherited anemia in which erythrocytes sickle under low oxygen conditions.

529. sieve plate

A large pore through which water enters echinoderms.

530. sieve tube

Linearly aligned specialized cells that form phloem in plants and transport nutrients.

531. spectrin

A protein that maintains flexibility of erythrocyte membranes.

532. sphygmomanometer

A device used to measure blood pressure.

533. stele

A cylinder of vascular tissues in the roots and stems of plants.

534. stenosis

Narrowing or constriction.

535. **stethoscope**

An device used to amplify sounds from inside the body.

536. **streptokinase**

An enzyme that digests blood clots used to treat coronary occlusion.

537. **stroke volume**

Volume of blood ejected by one ventricle during a single contraction.

538. **suberin**

A fatty substances in the walls of certain plant cells.

539. **systole**

Contraction.

540. **T lymphocyte**

A thymus-processed lymphocyte that can differentiate into several kinds of T cells.

541. **T wave**

A part of an electrocardiogram that occurs as ventricles repolarize.

542. **thrombin**

An enzyme that activates fibrinogen to fibrin in the blood clotting mechanism.

543. **thrombocyte**

A platelet.

544. **thrombus**

A blood clot that is stationary in a blood vessel wall.

545. **tissue plasminogen activator (tPA)**

A substance secreted by many tissues that activates plasminogen to plasmin.

546. **tissue thromboplastin**

A substance from injured tissue that initiates extrinsic blood clotting.

547. **torr**

A unit of pressure equal to that required to support a column of mercury 1 mm tall.

548. tracheid

A hollow woody cell with pointed ends specialized for water conduction.

549. transferrin

An iron-transport protein in plasma.

550. transpiration

The loss of water vapor from plants.

551. transpiration-cohesion theory

A theory that water rise in plants is due to transpiration, cohesion, and root pressure.

552. tricuspid

Having three points or cusps.

553. tunica

A layer or coat.

554. valve

A structure in a passageway that prevents reflux.

555. vascular

Pertaining to or full of vessels.

556. vascular bundle

Xylem and phloem enclosed in a cylindrical sheath, mainly in herbaceous plant stems.

557. vascular cambium

Cells capable of dividing to form new xylem and phloem within a vascular bundle.

558. vascular cylinder

A central region in dicot roots that contains bundles of vascular tissues, xylem and phloem.

559. vascular system

A transport system.

560. vein

A blood vessel that carries blood to the heart.

561. **vena cava**

A large vein that empties blood into the heart.

562. **venous**

Of a vein.

563. **ventricle**

A small cavity.

564. **venule**

A small vessel between capillaries and a vein.

565. **viscosity**

A fluid's tendency to resist flow.

566. **xylem**

The water conducting tissue of a plant.

Body Defenses

567. acquired immune deficiency syndrome (AIDS)

A viral disease that severely impairs immunity.

568. acquired immunity

Disease resistance obtained from another's antibodies.

569. active immunity

Disease resistance obtained by the immune system responding to a microorganism or a vaccine.

570. agglutinin

An antibody in an agglutination reaction.

571. agglutinogen

An antigen that elicits an agglutination reaction.

572. allergen

A substance capable of eliciting an allergic reaction.

573. allergy

Unusual sensitivity to a normally harmless concentration of substance.

574. anaphylaxis

A severe allergic reaction to a substance to which one has been previously sensitized.

575. antibody

A protein released in response to an antigen that can inactivate the antigen.

576. antigen

A substance that elicits a response from the immune system.

577. artificially acquired active immunity

Disease resistance obtained by stimulating the immune system with a vaccine.

578. artificially acquired passive immunity

Temporary disease resistance obtained by receiving another's antibodies.

579. B lymphocyte

A lymphocyte that produces plasma cells, which in turn produce antibodies.

580. bacterial antagonism

Inhibition of growth of one bacterial species by another species.

581. bursa of Fabricius

A structure found in birds where B lymphocyte differentiation was first identified.

582. cell-mediated immunity

Disease resistance involving direct destruction of antigenic cells.

583. chemotaxis

The act of a chemical stimuli to attract or repel; a process that causes some leukocytes to migrate to sites of injury.

584. clonal selection theory

Explanation of how lymphocytes are sensitized to a certain antigen and how immune tolerance for self arises.

585. clone

A group of genetically like cells derived from a single parent cell.

586. complement

A group of plasma enzymes that catalyze a sequence of reactions against many different kinds of foreign matter.

587. cuticle

The outer layer of skin, especially around nails.

588. dermis

A thick skin layer underlying the epidermis.

589. eleidin

A keratin precursor found in the stratum lucidum.

590. epidermis

Outer skin layer consisting of epithelium.

591. fever

An abnormally high body temperature.

592. heterogeneity

Diversity.

593. histamine

A derivative of the amino acid histidine released by injured cells that causes vasodilation and bronchial constriction.

594. humoral immunity

Disease resistance produced by antibodies.

595. hybridoma

A cell made by fusing parts of two cells.

596. hypersensitivity

Abnormal reaction to a substance, as occurs in allergy and certain other immune reactions.

597. IgA

An immunoglobulin in secretions.

598. IgD

An immunoglobulin of unknown function.

599. IgE

An immunoglobulin responsible for allergic responses.

600. IgG

An immunoglobulin in blood and primarily responsible for resisting infection.

601. IgM

A multiunit immunoglobulin most abundant early in an immune response.

602. immune

Disease resistant.

603. immunity

A state of disease resistance.

604. immunization

Use of a vaccine or other procedure to create immunity.

605. immunodeficiency

An absence or lack of a normal immune function.

606. immunoglobulin

A protein that can bind with a foreign substance; antibody that binds with an antigen.

607. immunology

The study of immunity and immune reactions.

608. immunosuppression

A procedure used to lessen an immune response.

609. immunotoxin

An antibody bound to a toxic drug.

610. inflammation

Localized response to tissue injury, usually involving redness, swelling, increased temperature, and pain.

611. interferon

A protein released by virally-infected cells that causes adjacent cells to make an antiviral protein.

612. interleukin

A substance that facilitates or enhances an immune reaction.

613. kinin

A substance that stimulates events in the inflammatory process.

614. lanugo

Fine hair on the skin of a fetus.

615. lymphokine

A substance that stimulates activity of lymphocytes.

616. macrophage

A large phagocytic cell in connective tissue.

617. melanin

A dark brown pigment of hair and skin.

618. naturally acquired active immunity

Immunity produced by having a disease.

619. naturally acquired passive immunity

Immunity based on antibodies transferred across the placenta or in breast milk.

620. papilla

A nipple-shaped projection.

621. passive immunity

Temporary disease resistance derived from antibodies from another organism (human or other).

622. phagocytosis

Engulfment into a vacuole and digestion by a scavenger cell.

623. pilus

A hair.

624. plasma cell

An antibody-producing cell derived from a B lymphocyte.

625. pus

A product of inflammation consisting of debris from dead leukocytes and microorganisms.

626. sebum

A substance containing oils and epithelial cell debris from sebaceous glands.

627. sensitization

Rendering a lymphocyte sensitive to a foreign substance.

628. T cell

T lymphocyte.

629. theory of immune surveillance

A possible way the body finds malignant cells and destroys them.

Excretion and Fluid Regulation

630. acid-base balance

Maintenance of body fluid pH withing a normal range.

631. ammonotelic excretion

Excretion of ammonia as the end product of nitrogen metabolism.

632. Bowman's capsule

Glomerular capsule of kidney.

633. calyx

Cup-shaped cavity or organ.

634. clearance

Rate at which the kidneys can remove a substance from the blood.

635. collecting duct

One of many ducts that receive filtrate from kidney tubules.

636. countercurrent mechanism

A process in which a system's outflow affects its inflow.

637. electrolyte

Substance that ionizes and conducts electricity.

638. excretion

Elimination of a waste product.

639. fenestrated

Having one or more openings; having windows.

640. fluid regulation

Maintenance of body fluid volumes within normal ranges.

641. glomerulus

A capillary tuft surrounded by a glomerular capsule.

642. loop of Henle

U-shaped segment of a nephron where sodium chloride becomes concentrated in peritubular fluid.

643. Malphighian tubule

An excretory organ in insects.

644. mesonephros

Temporary embryonic kidney.

645. metanephridium

An advanced kind of invertebrate excretory structure.

646. metanephros

The embryonic kidney from which the mammalian functional kidney is derived.

647. nephridium

An excretory organ found in many invertebrates, especially segmented worms.

648. nephron

The functional unit of a mammalian kidney.

649. net filtration pressure

Pressure pushing materials out of a blood vessel.

650. peritubular

Surrounding a tubule, as in the kidney.

651. pronephros

The most primitive embyronic kidney among vertebrates.

652. renin

Kidney secretion that activates angiotensinogen to angiotensin I.

653. renin-angiotensin mechanism

A mechanism that increases blood pressure and blood volume when either falls below normal.

654. thirst

Desire for water or other fluid.

655. ureotelic excretion

Excreting nitrogen mainly as urea.

656. ureter

Tube through which urine flows from a kidney to the urinary bladder.

657. urethra

Tube through which urine flows from the urinary bladder outside the body.

658. uricotelic excretion

Excreting nitrogen mainly as uric acid.

659. urinary bladder

Distensible sac where urine is stored.

660. water balance

A state in which water intake and water output are equal.

Neurons and Sensory Systems

661. **accommodation**

   Adjustment of the focal distance of the eyes to see close objects clearly.

662. **acetylcholine**

   A neurotransmitter released by many axons, especially those that control skeletal muscles.

663. **action potential**

   Wave of change in electrical potential across the cell membrane of an excited cell; impulse.

664. **adaptation**

   Decrease in excitability of sensory receptors after a period of continuous, constant intensity stimulation.

665. **adenohypophysis**

   Anterior pituitary gland.

666. **adrenergic**

   Concerning a neuron that releases norepinephrine (adrenalin).

667. **afferent**

   Leading toward.

668. **Alzheimer's disease**

   A degenerative neurological disorder associated with memory loss and behavioral changes.

669. **amplitude**

   Signal strength; the intensity of a sound.

670. **ampulla**

   A dilation in a passageway.

671. **anesthetic**

   Agent that produces temporary loss of sensation.

672. **aqueous**

   Watery.

673. arachnoid

Spiderlike, as the delicate middle meninges that covers the brain and spinal cord.

674. arbor vitae

Treelike pattern of cerebellar white matter.

675. association neuron

A neuron that relays impulses from sensory to motor neurons, especially in the spinal cord.

676. astigmatism

Blurred vision due to irregular curvature of one or more refractive surfaces in the eye.

677. astrocyte

A star-shaped neuroglial cell in the central nervous system.

678. auditory

Of hearing.

679. autonomic nervous system

A nervous system component that regulates internal organ functions and involuntary processes.

680. axon

The part of a neuron that typically carries impulses away from the cell body toward another neuron.

681. axon terminal

The end of an axon from which neurotransmitter is released.

682. basilar membrane

Inner ear membrane associated with sound receptors.

683. bilateral

Relating to both sides of the body; having left and right sides.

684. binocular vision

Sight using two eyes that allows perception of three-dimensionality.

685. biofeedback

Use of signals indicating levels of autonomic process to control the process.

686. blind spot

Region of the retina that lacks light receptors where optic nerve fibers leave the eye.

687. brachial

Of the arm.

688. brain stem

Brain parts that relay impulses to and from the cerebrum, cerebellum, and other brain structures.

689. brain wave

An electrical signal detected on the scalp that represents brain activity.

690. Broca's motor speech area

A functional area of the cerebrum where thoughts are translated into speech.

691. cataract

Opacity of the eye lens.

692. catecholamine

A class of amines that act as chemical messengers; dopamine, epinephrine, and norepinephrine.

693. central nervous system (CNS)

Brain and spinal cord.

694. cephalization

The concentration of sensory organs and nervous tissue at the head end of an animal.

695. cerebellum

A brain component behind the cerebrum and above the pons concerned with the coordination of movements.

696. cerebrospinal fluid

A clear fluid in spaces within and around the central nervous system.

697. cerebrovascular

Relating to blood vessels of the brain.

698. cerebrum

The largest brain component; responds to sensory impulses and carries out mental processes.

699. cerumen

Earwax.

700. chemoreceptor

A receptor that responds to certain chemical substances.

701. chiasma

A site at which fibers cross over.

702. cholinergic

Relating to a neuron whose terminals release acetylcholine.

703. cholinesterase

An enzyme that degrades acetylcholine.

704. cholinesterase inhibitor

A substance that blocks cholinesterase action.

705. choroid

Vascular middle layer of the eyeball.

706. ciliary body

Anteriormost part of the choroid layer, which contains ciliary muscles that participate in accommodation.

707. circle of Willis

A ring of blood vessels at the base of the brain by which blood reaches alternate circulatory pathways in the brain.

708. cochlea

Snail-shaped bony part of the inner ear.

709. conduction deafness

Loss of hearing due to impaired transmission of vibrations to sound receptors.

710. cone

A light receptor that responds to a certain color.

711. conjunctiva

A mucous membrane lining the eyelids and covering the anterior eyeball surface.

712. consciousness

Awareness of signals from the sense organs.

713. convergence

Coming together.

714. cornea

A transparent part of the anterior eye surface.

715. corpus callosum

An aggregation of myelinated neural fibers that carry impulses between the two cerebral hemispheres.

716. cortex

Outer portion of an organ; literally, bark.

717. craniosacral

Of the cranial and sacral regions.

718. cupula

A domelike cup-shaped structure.

719. decibel

A unit on the logarithmic scale of sound intensity.

720. dendrite

A cytoplasmic process of a neuron that commonly receives signals from other neurons.

721. dura mater

The tough outermost meninges that surrounds the brain, spinal cord, and other meninges.

722. efferent

Leading away from.

723. emotion

A state of feeling; the affective aspect of consciousness.

724. exteroceptor

A sensory receptor that detects environmental changes.

725. fovea centralis

A pit in the retina that contains only cone receptors; region of greatest visual acuity.

726. frequency

Number of occurrences of an event in a give period, such as vibrations per second of sound waves.

727. ganglion

An aggregation of cell bodies in the peripheral nervous system.

728. glaucoma

A disorder in which aqueous humor accumulates and exerts excessive intraocular pressure.

729. gray matter

Unmyelinated tissue (mainly cell bodies) in the central nervous system.

730. gustatory

Pertaining to the sense of taste.

731. gyrus

A fold, such as a convolution of the cerebrum.

732. habit-forming

A property that causes some people to make great effort to obtain a drug.

733. habituation

Gradual adaptation to a continuing stimulus.

734. hair cell

A kind of sensory receptor with an easily stimulated thin projection.

735. hippocampus

A part of the limbic system in the temporal region and concerned with emotion and memory.

736. hypophysis

The pituitary gland.

737. hypothalamus

A part of the brain that connects and serves both the nervous and endocrine systems.

738. innate

Already present at birth.

739. innervation

Nerve supply.

740. interneuron

A neuron that relays impulses between a sensory and a motor neuron, typically found in the spinal cord.

741. iris

Muscular diaphragm anterior to the eye lens that regulates the amount of light entering the eye.

742. isthmus

Constriction; neck.

743. kinesthetic

Concerned with sensing movement.

744. labyrinth

Maze.

745. lacrimal

Of tears.

746. lateral inhibition

Suppression of hair cells adjacent to those stimulated, which makes signals from stimulated receptors clearer.

747. lateral line organ

Sensory receptors found on the body surface of certain fish that allows them to detect objects in the water.

748. lateralization

Difference in function of the two sides of a bilateral structure, especially the cerebrum.

749. law of adequate stimulus

A law stating that a receptor responds only if it receives a sufficiently strong stimulus.

750. learning

A behavioral change in response to external stimuli.

751. lens

Transparent, biconcave structure behind the iris that focuses eye on far or near objects by changing shape.

752. limbic system

A part of the brain mainly associated with emotions.

753. lysozyme

An enzyme in tears that can destroy microbes.

754. macula

An inner ear structure that contains sensory receptors for static equilibrium.

755. mechanoreceptor

A receptor that responds to mechanical pressure.

756. medulla

The inner core of an organ.

757. medulla oblongata

A part of the brain continuous with the spinal cord.

758. membrane potential

An electrical potential (potential difference between) the inside and outside of a membrane.

759. memory

A process of storing and recalling previous experiences.

760. Meniere's disease

A semicircular canal inflammation that leads to sight, hearing, and balance disorders.

761. meninges

Membranes that surround the brain and spinal cord.

762. mesencephalon

A region of the embryonic brain near the middle of the brain.

763. microglia

Small supporting cells of the central nervous system.

764. mixed nerve

A nerve with both sensory and motor fibers.

765. modiolus

A structure that supports the cochlea.

766. monosynaptic

Having a single junction between neurons.

767. motor neuron

A neuron that carries impulses toward a muscle or gland.

768. myelin

An insulating substance deposited around axons.

769. neocortex

The most recently evolved part of the cerebrum.

770. nerve

A bundle of axons covered with a connective tissue sheaths.

771. nerve net

A simple nervous system found in cnidarians.

772. neural crest

Cells that give rise to sensory neurons, adrenal medulla, and the autonomic nervous system.

773. neural oscillator

A process that controls movements or other behaviors.

774. neural tube

A tube formed by invagination of ectoderm that produces to the nervous system.

775. neurilemma

The Schwann cell membrane.

776. neurofibrillary tangles

Masses of disorderly neural fibers in the brains of Alzheimer patients.

777. neuroglia

Supporting cells of the nervous system.

778. neurotransmitter

A chemical substance from one neuron that transmits a signal to another neuron at a synapse.

779. neurulation

Formation of a neural tube as a part of embryonic development.

780. Nissl granule

Granules containing RNA and endoplasmic reticulum in nerve cell bodies and dendrites.

781. nociceptor

A receptor that responds specifically to painful stimuli.

782. node of Ranvier

A gap in an axon's myelin sheath.

783. noradrenalin

Norepinephrine.

784. norepinephrine

A neurotransmitter of the sympathetic division of the autonomic nervous system and of some brain neurons.

785. olfactory

Of the sense of smell.

786. opsin

A protein that combines with retinine in the retina.

787. optic chiasma

A site anterior to the pituitary where medial fibers of each optic nerve cross from one side of the body to the other.

788. optic disk

Region of the retina lacking receptors where optic nerve fibers leave the eyeball; blind spot.

789. otolith

A small calcium carbonate particle in a receptor for static equilibrium; an ear stone.

790. oval window

A membrane-covered opening from the middle to the inner ear across which the stapes transmits vibrations.

791. oxytocin

A hormone from the hypothalamus that stimulates uterine contractions and milk let down.

792. Pacinian corpuscle

A receptor that responds to pressure.

793. parasympathetic division

Autonomic component that accelerates digestion and other functions not essential to a response to stress.

794. parathormone

A hormone from the parathyorid gland that decreases blood calcium.

795. parathyroid glands

Glands imbedded in the thyorid gland.

796. Parkinson's disease

Muscle rigidity and tremors due to a dopamine deficiency in the brain.

797. pattern generator

A control mechanism for repetitive movements.

798. perception

Conscious interpretation of information from sensory receptors.

799. perikaryon

Substance around a nucleus; the cell body of a neuron.

800. photon

The smallest unit of light energy.

801. photoreceptor

A receptor that responds to light.

802. physiological dependence

The need (or perceived need) for a drug if one is to prevent withdrawal symptoms.

803. pia mater

A delicate membrane on the surface of the brain and spinal cord.

804. pineal gland

A gland between the cerebral hemispheres associated with regulating circadian rhythms.

805. pinna

The flap-like part of the ear projecting from the head.

806. pitch

A sound quality determined by vibration frequency.

807. plexus

A network of nerves or blood vessels.

808. pons

A part of the brain stem associated with the cerebellum.

809. postganglionic

Concerning a neuron that receives signals across a synapse in a ganglion; second neuron in an autonomic pathway.

810. postsynaptic

Referring to a neuron that receives a neurotransmitter at a synapse.

811. preganglionic

Referring to an neuron that sends a signal through a ganglion.

812. presynaptic

Referring to a neuron that releases a neurotransmitter at a synapse.

813. principle of forward conduction

A rule that signals travel along axons toward the next neuron in a pathway.

814. proprioceptor

Any sensory receptor in a muscle, joint, or tendon that detects position or movement.

815. Purkinje cell

A cerebellar neuron that synapses with a very large number of other cells.

816. rapid-eye-movement (REM) sleep

Sleep interval in which the eyeballs move and the EEG resembles wakefulness; paradoxical sleep.

817. rarefaction

Decreased density.

818. reflex

An automatic involuntary response to a stimulus.

819. refraction

Bending of light rays as they pass from a medium of one density to a medium of a different density.

820. response

An action elicited by a stimulus.

821. reticular activating system (RAS)

A brain stem structure involved in maintaining consciousness.

822. retina

The innermost layer of the eye, which contains light receptors.

823. retinene

A carotenoid pigment that binds to opsin.

824. rhodopsin

A light-sensitive protein found in rods of the retina.

825. rod

A receptor in the retina that responds to different intensities of light but not to color.

826. root

A base or foundation.

827. round window

A membrane-covered opening between the middle and inner ear.

828. saltatory

Leaping.

829. Schwann cell

A myelin producing cell in the peripheral nervous system.

830. semicircular canal

One of three pairs of fluid-filled inner ear passageways that detect head movements.

831. sensation

An impression conveyed.

832. sensory

Concerned with sensation.

833. serotonin

A substance secreted as a brain neurotransmitter and a gut hormone.

834. soma

Cell body.

835. somatic

Of the body.

836. static equilibrium

Maintenance of balance while the head is stationary.

837. statocyst

An organ of balance in a crayfish.

838. stimulus

An event that typically elicits a response.

839. stretch reflex

Muscle contraction following stimulation of stretch receptors in a muscle or its tendon.

840. subconscious

Partially conscious.

841. sympathetic chain ganglion

An aggregation of the cell bodies of postsynaptic neurons of the sympathetic division.

842. sympathetic division

The part of the autonomic nervous system that can respond to stressful situations.

843. synapse

A junction where a signal passes from one neuron to the next in a pathway, usually by neurotransmitter diffusion.

844. synaptic

Of a synapse.

845. target cell

A cell that can respond to a certain hormone.

846. taste bud

A structure on the tongue containing taste receptor cells.

847. tear

A fluid released from tear glands.

848. thalamus

Subcortical gray matter near the anterior end of the brain stem.

849. thermoreceptor

A receptor that detects temperature changes.

850. thyroid gland

A gland in the throat that produces metabolism regulating hormones.

851. tolerance

The requirement for larger doses of a substance to produce an effect.

852. transducin

Enzyme involved in visual process.

853. transduction

The conversion of a signal from one type to another, such as from chemical to electrical.

854. tympanic membrane

Eardrum; membrane between the external and middle ear.

855. utricle

Large chamber in the ear vestibule that contains receptors for equilibrium.

856. vestibule

A space or cavity near the entrance of a canal.

857. visceroceptor

A receptor in or near an internal organ.

858. visual area

A region of the occipital lobe of the cerebrum that receives and processes signals from the retina.

859. vitreous

Glassy.

860. Wallerian degeneration

Axon disintegration in an injured neuron distal to the point of injury.

861. white matter

Myelinated nerve fibers of the central nervous system.

862. withdrawal reflex

A reflex that leads to flexion and removal of a limb from a painful stimulus.

Support and Movement

863. abduction

Motion of a body part away from the midline.

864. Achilles tendon

A tendon in the heel.

865. actin

A contractile protein.

866. action

A movement produced by one or more muscles.

867. agonist

The prime mover among a group of muscles.

868. antagonist

A muscle that opposes an agonist.

869. articulation

A joint.

870. atrophy

A decrease in size, usually accompanied with reduced function.

871. ball-and-socket joint

A joint with a ball-shaped articular surface of one bone fiting into a socket-shaped articular surface of another.

872. bursa

A sac containing synovial fluid located at pressure points or near joints.

873. calcification

Depositing of calcium salts in an organic matrix.

874. callus

A thickened area.

875. carpal

A wrist bone.

876. **cartilage**

A firm, resilient, flexible connective tissue.

877. **clavicle**

Bone that extends from the sternum to the scapula.

878. **coccyx**

The caudal end of the spinal column; tail bone.

879. **concha**

One of several shell-shaped bones in the nasal cavity.

880. **contractile protein**

A protein that acts in shortening a muscle or causing it to develop tension.

881. **contractility**

The ability to develop tension or shorten.

882. **contraction cycle**

Repetitive sliding actions of actin and myosin in a muscle filament as it develops tension.

883. **coxal bone**

A bone that forms half of the pelvic girdle.

884. **cranium**

Skull bones that surround the brain.

885. **creatine phosphate**

A molecule that accounts for limited energy storage in muscle.

886. **cross-bridge**

The end of a myosin filament bound to actin during muscle contraction.

887. **diaphragm**

A thin partition, as is formed by a muscle between the thoracic and abdominal cavities.

888. **excitation-contraction coupling**

The means by which neural signals excite muscle cells and cause contraction.

889. extensibility

Ability to be extended or stretched.

890. fascia

Fibrous connective tissue sheath around muscles and beneath skin.

891. fatigue

Loss of power for a short time.

892. fibula

Lateral leg bone found between the knee and ankle.

893. flexion

A movement that decreases the angle between two bones.

894. fontanel

A membranous non-bony region between cranial bones in an infant.

895. fracture

A break, such as in a bone.

896. frontal

Of the forehead.

897. fulcrum

Fixed point about which a lever produces movement.

898. humerus

The bone of the upper arm.

899. ilium

The posteriolateral bone of the pelvis.

900. incus

Anvil; the ear bone that receives vibrations from the malleus.

901. insertion

The most moveable attachment of a muscle to a bone.

902. ischium

A posterior bone of the pelvic girdle.

903. isometric

Having the same length.

904. joint

A connection between two or more bones.

905. latent period

In muscle physiology, a time between the application of a stimulus and the beginning of muscle contraction.

906. ligament

Cord of fibrous connective tissue that attaches bones to each other.

907. malleus

Hammer; the outermost bone of the middle ear.

908. marrow

Fatty substance found in a marrow cavity.

909. maxilla

The upper bone bone that contains sockets for upper teeth.

910. mesenchyme

Embryonic mesoderm.

911. metacarpal

One of five bones in the palm of the hand.

912. metatarsal

One of five long bones in the foot.

913. motor end plate

A portion of sarcolemma lying beneath nerve endings.

914. motor unit

A motor neuron and the muscle fibers it innervates.

915. myofibril

A contractile fiber in a muscle cell.

916. myofilament

A component of a myofibril consisting of one or more protein molecules.

917. myoglobin

A pigmented protein that binds oxygen in muscle tissue.

918. myoneural junction

The structure at which nerve and muscle tissues meet and impulses are relayed.

919. myosin

A protein that comprises thick filaments of a myofibril.

920. myotome

A block of mesoderm from which muscle arises.

921. neuromuscular

Concerning the association between the nervous and muscular systems.

922. occipital

Of or near the back of the head.

923. origin

The least movable attachment of a muscle to a bone.

924. ossification

Mineral deposition in the process of bone formation.

925. osteon

The cells, matrix, and passages that make up a unit of compact bone.

926. oxygen debt

The quantity of oxygen required to oxidize metabolites produced anaerobically during strenuous activity.

927. patella

A small bone that forms the kneecap.

928. phalanges

Small, slightly elongated bones of the fingers and toes.

929. plantar

Of the sole of the foot.

930. plasticity

Able to be molded or formed; changeable.

931. prime mover

A muscle that is the most direct cause of a particular movement.

932. pubis

An anterior bone of the pelvic girdle.

933. radius

The smaller of the forearm bones.

934. raphe

A seamlike ridge, usually where two structures have fused.

935. rickets

A failure of bones to harden in childhood because of a calcium deficiency.

936. rotation

Motion of a part about its own axis.

937. sacrum

A bone formed from the fusion of five vertebrae that articulates posteriorly with the pelvic girdle.

938. sarcomere

The contractile unit of skeletal muscle.

939. sarcoplasm

The protoplasmic, nonfibrillar substance of a muscle cell.

940. sarcoplasmic reticulum

A vesicular network associated with myofibrils of a striated muscle cell.

941. scapula

A large bone lateral to the vertebrae in the shoulder area.

942. sclerotome

An embryonic tissue from which vertebrae and ribs develop.

943. sinus

A cavity or recess.

944. sliding filament theory

An explanation of how myofilaments move with respect to each other during muscle contraction.

945. smooth muscle

A type of muscle located in the walls of hollow organs and blood vessels.

946. spicule

A needle-shaped structure.

947. stapes

The middle ear bone that transmits vibrations to the oval window.

948. striation

Stripe.

949. sulcus

A furrow or groove.

950. summation

Addition, as in effects of multiple stimuli to a muscle.

951. suture

An fibrous joint at which no movement occurs.

952. symphysis

A cartilaginous, slightly moveable joint.

953. syncytium

A group of cells that lack membranes to separate them.

954. synergist

A muscle that works with a prime mover.

955. synovial joint

A freely movable joint.

956. tarsal

Of the foot bones.

957. temporal

Time-related; a brain lobe where auditory and olfactory areas are located.

958. tendon

A fibrous connective tissue cord that holds a muscle to a bone.

959. tension

A pulling force.

960. tetanus

A sustained contraction maintained by repeated muscle stimulation.

961. tibia

Larger, more medial weight-bearing bone of the leg.

962. tonus

A slight continuous muscle contraction.

963. transverse (T) tubule

Crosswise tubule in skeletal muscle myofibrils that carries signals from the sarcolemma to the myofibrils.

964. trochlea

Pulley.

965. tropomyosin

A muscle protein that alters the actin configuration so that contraction can occur.

966. troponin

A muscle protein that binds to tropomyosin causing it to alter the configuration of actin.

967. twitch

A muscle response to a single stimulus.

968. ulna

Larger of the forearm bones.

969. vertebra

Back bone.

970. vomer

A shovel-shaped bone that forms the nasal septum.

971. zygomatic

Of the cheek bone.

Hormones

972. adrenal

Above the kidney; a gland lying superior to the kidney.

973. adrenalin

A hormone secreted by the adrenal glands.

974. adrenocorticotropic hormone

A hormone that stimulates the adrenal cortex to secrete hormones.

975. aldosterone

An adrenocortical hormone that increases reabsorption of sodium.

976. anabolic steroid

A synthetic hormone that increases muscle size.

977. androgen

A molecule with male hormone activity.

978. antidiuretic hormone (ADH)

A hypothalamic hormone stored in the posterior pituitary gland that stimulates water conservation by the kidneys.

979. calcitonin

A hormone that lowers blood calcium.

980. corpus allata

Endocrine gland of insects that secretes juvenile hormone.

981. corpus cardica

Aggregate of nerve cells in insects that releases brain hormones.

982. cortisol

An adrenocortical hormone that helps regulate carbohydrate metabolism and counteracts inflammation.

983. endocrine

Concerning a ductless gland.

984. epinephrine

Main hormone from the adrenal medulla.

985. follicle-stimulating hormone (FSH)

A hormone that stimulates maturation of ova and sperm.

986. gastrin

A hormone from the stomach lining that circulates in the blood and stimulates HCl secretion.

987. glucagon

A hormone that raises blood glucose.

988. glucocorticoid

A hormone that helps to regulate carbohydrate metabolism.

989. hormone

A regulatory substance from an endocrine cell that is transported in the blood to its target cells.

990. insulin

A hormone from the pancreas that causes cells to take in glucose and stimulates protein synthesis.

991. islet of Langerhans

Cluster of hormone-secreting cells in the pancreas.

992. luteinizing hormone (LH)

A hormone that helps to cause ovulation and other reproductive processes.

993. mineralocorticoid

A hormone that regulates mineral metabolism.

994. negative feedback

A control system in which a system's output inhibits activity of the system.

995. neurohypophysis

The posterior, neural part of the pituitary gland.

996. positive feedback

Output from a system that accelerates change in the system.

997. progesterone

A hormone that helps to maintain pregnancy.

998. prolactin

A hormone that stimulates milk secretion.

999. prostaglandin

A substance derived from the fatty acid arachidonic acid that acts over short distances as a chemical messenger.

1000. puberty

A period during which sexual maturity is achieved.

1001. Rathke's pouch

An embryonic structure from which the anterior pituitary develops.

1002. relaxin

A hormone from the corpus luteum of pregnancy.

1003. somatostatin

Growth-hormone inhibiting hormone.

1004. somatotropin

Growth hormone.

1005. stress

A condition produced by a variety of injurious agents that affects many body systems.

1006. stressor

An agent or event that produces stress.

1007. testosterone

A male hormone.

1008. thymosin

A hormone secreted by the thymus gland.

1009. thymus gland

A gland that processes and activates T lymphocytes before it regresses during puberty.

1010. thyroid-stimulating hormone (TSH)

A hormone that stimulates hormone secretion by the thyroid gland.

1011. thyroxine

Hormone produced by the thyroid gland that regulates metabolic rate.

1012. tract

A bundle of myelinated neurons in the brain or spinal cord.

1013. tropic

Influencing another organ or process.

Animal Behavior

1014. aggression

An attitude of hostility, which may or may not include an attack.

1015. biological clock

A mechanism that enables organisms to respond to time related changes in the environment.

1016. browser

An animal that feeds above the ground on trees and shrubs.

1017. circadian rhythm

A biological cycle that is repeated about every 24 hours.

1018. cleaning symbiosis

A mutualistic relationship in which both organisms gain, one by cleaning and the other by being cleaned.

1019. competition

Interactions in which two organisms or two species limit each other's supply of food, shelter, or mates.

1020. conditioning

A kind of learning in which a stimulus and response become associated.

1021. courtship ritual

A pattern of activities preceding mating that are needed for mating to occur.

1022. dominance hierarchy

A social pattern in which a few animals dominate the majority.

1023. entrainment

The following of environmental cues in a time-related cycle.

1024. estivation

A period of inactivity during hot weather.

1025. ethology

The study of animal behavior.

1026. grazer

An animal that feeds on grasses and other plants near the ground.

1027. hibernation

A period of greatly reduced metabolism during cold weather.

1028. imprinting

A special kind of learning that occurs during a brief sensitive time early in an animal's life.

1029. inclusive fitness

Fitness attributed to characteristics possessed by ancestors and close relatives.

1030. individual fitness

The extent to which an individual's genes survive in its offspring.

1031. individual space

Space immediately surrounding an animal that it usually does not permit another animal to enter.

1032. kin selection

Tendency to protect close relatives, thereby fostering survival of one's own genes in others.

1033. latent learning

Learning having no immediate and apparent use.

1034. learned behavior

Behavior that results from experience.

1035. mimicry

A resemblance of one organism to another that can defend itself against a common predator.

1036. operant conditoning

Learning in which an animal is encouraged to repeat a behavior by being rewarded for performing it.

1037. pair bond

A special relationship between mating pairs that persists beyond the act of mating; seen in some birds and mammals.

1038. pheromone

A substance released by some organisms into the environment to communicate with others of the same species.

1039. photoperiodism

The dependence of flowering on the length of day or night.

1040. releaser

A stimulus that initiates a given behavior.

1041. sociobiology

The study of the biological basis of behavior.

1042. stereotyped behavior

Repetitive behavior that tends to follow a particular stimulus.

1043. stridulation

Audible vibrations produced by insects rubbing appendages together.

1044. taxis

Movement of a protist or animal toward or away from a stimulus.

1045. territoriality

Establishment and defense of a nesting, breeding, or feeding site in which others of same species are driven away.

1046. territory

An area actively defended by an individual or group that inhabits it.

1047. zeitgeber

An environmental stimulus that serves to set an organism's biological clock.

Animal Reproduction

1048. acrosome

Dense anterior part of a sperm that contains enzymes needed to penetrate an ovum.

1049. alternation of generations

A reproductive cycle involving gametophyte and sporophyte generations of plants.

1050. amenorrhea

Absence of menstruation.

1051. areola

Pigmented region around the nipple in a mammary gland.

1052. artificial insemination

Method for introducing sperm into the female reproductive tract without sexual intercourse.

1053. asexual reproduction

Reproduction not involving the union of gametes.

1054. Bartholin's gland

Vulvovaginal gland.

1055. budding

Asexual reproduction in which a new organism or cell pinches of from the parent.

1056. cervix

Neck; narrow end of the uterus adjacent to the vagina.

1057. clitoris

Small erectile organ homologous to the penis located in the anterior vulva.

1058. colostrum

First fluid from a mammary gland after childbirth.

1059. corpus albicans

A scar on an ovary that remains after the degeneration of a corpus luteum; literally, white body.

1060. corpus luteum

Follicular cells that produce hormones after ovulation; literally, yellow body.

1061. diploid

Having paired chromosomes.

1062. ejaculation

Ejection of semen.

1063. erection

Rigid state of the penis.

1064. estrogen

A kind of female hormone that stimulates developemtn of sex organs and secondary sexual characteristics.

1065. external fertilization

The union of sperm and eggs outside parental bodies.

1066. fallopian tube

Uterine tube.

1067. female pronucleus

Nucleus of an ovum that fuses with the nucleus of a sperm.

1068. fertilized ovum

An ovum already penetrated by a sperm.

1069. fetus

Unborn child from two months development to birth.

1070. fission

Asexual reproduction in which a cell divides into two.

1071. gamete

Haploid cell; an ovum or sperm.

1072. gametogenesis

The process of forming gametes.

1073. genitalia

Genital, or sex, organs.

1074. glans

The cone-shaped tip of the penis or clitoris.

1075. gonad

An ovum or sperm.

1076. gonorrhea

A sexually transmitted disease caused by Neisseria gonorrhoeae.

1077. graafian follicle

An ovarian follicle.

1078. implantation

Attachment of an embryo to the endometrium of the uterus.

1079. internal fertilization

The union of egg and sperm within a parental, usually female, body.

1080. isogamy

Having structurally identical male and female gametes.

1081. labia

Lip-shaped structures.

1082. lactation

Synthesis and secretion of milk.

1083. lactiferous

Making or conveying milk.

1084. luteum

Yellow.

1085. male pronucleus

Nucleus of a sperm that has penetrated an ovum.

1086. mammary gland

Gland that synthesizes and secretes milk.

1087. meiosis

Cell division that gives rise to haploid cells.

1088. menarche

Onset of menstruation during puberty.

1089. menopause

Cessation of menstruation.

1090. menstrual cycle

Repetitive sequence of events involving ovulation and preparation of the uterus for implantation.

1091. menstruation

Periodic discharge of blood, tissue debris, and fluid from the uterus.

1092. mons pubis

Fatty area covered with pubic hair.

1093. oocyte

Cell that gives rise to an ovum.

1094. oogenesis

Process of producing an ovum.

1095. oogonia

Female germ cells.

1096. orgasm

An intense pleasurable culmination related to sexual intercourse.

1097. ovary

A female gonad.

1098. oviduct

Uterine tube or Fallopian tube.

1099. oviparous

Pertaining to an animal that produces eggs that develop outside the mother's body.

1100. ovoviviparous

Pertaining to an animal that produces eggs that develop inside the mother's body.

1101. ovulation

Sudden expulsion of an ovum from a follicle.

1102. ovum

Female gamete.

1103. Pap smear

A cell sample from the cervix examined microscopically to detect cervical cancer.

1104. parturition

Childbirth.

1105. penis

Male copulatory organ.

1106. placenta

Structure attached to the uterine wall that provides nutrients and removes wastes for a developing fetus.

1107. polar body

A nonfunctional cell from unequal meiotic divisions during animal oogenesis.

1108. postovulatory

After ovulation.

1109. preovulatory

Before ovulation.

1110. prepuce

Foreskin of the penis.

1111. primary follicle

An early, immature stage of an ovarian follicle.

1112. primordial

Original or primitive.

1113. proliferative phase

A part of the menstrual cycle in which endometriual cells divide and increase in number.

1114. reproductive efficiency

The ratio of survivors to the total number of eggs produced.

1115. scrotum

A pouch in which the testes are located.

1116. semen

A fluid containing sperm and secretions from the male reproductive glands.

1117. seminal vesicle

A convoluted saclike organ near the ductus deferens that adds secretions to semem.

1118. seminiferous tubule

A coiled tubule within a testis in which sperm are produced.

1119. Sertoli's cell

Cell type in seminiferous tubules that nourishes developing sperm.

1120. sexual dimorphism

The presence of structural differences in males and females.

1121. sexual reproduction

Reproduction involving the production and union of gametes.

1122. sperm

A male gamete.

1123. spermatic cord

A cord extending from the scrotum to the inguinal ligament that includes ductus deferens, blood vessels, and nerves.

1124. spermatid

An immature spermatozoan.

1125. spermatocyte

A cell from which sperm develop.

1126. spermatogenesis

The process of sperm formation.

1127. spermatogonium

Undifferentiated male germ cells from which sperm-producing cells arise.

1128. spermatozoan

Male gamete, sperm.

1129. spermiogenesis

Maturation of spermatids to become sperm.

1130. sterility

Inability to produce offspring.

1131. stroma

Fibrous connective tissue framework that supports an organ.

1132. synapsis

Alignment of homologous pairs of chromosomes.

1133. syngamy

Fusion of gametes in fertilization.

1134. syphilis

A sexually transmitted disease caused by the spirochete Treponema pallidum.

1135. testis

A male gonad.

1136. tetrad

Four copies of the same chromosome temporarily attached to each other.

1137. theca

Sheath, such as that covering an ovarian follicle.

1138. umbilicus

Site where umbilical cord was attached to fetus; navel.

1139. uterine

Of the uterus.

1140. uterine tube

A tube between an ovary and the uterus.

1141. uterus

A hollow, pearshaped organ where a fetus develops.

1142. vagina

Passageway from the uterus.

1143. vas deferens

A duct between the epididymis and the ejaculatory duct; ductus deferens.

1144. viviparous

Producing living young, as opposed to laying eggs.

1145. zona pellucida

Translucent membrane around an oocyte in an ovarian follicle.

1146. zoospore

A flagellated, motile reproductive cell involved in asexual reproduction in some algae and fungi.

1147. zygote

Single cell resulting from a union of ovum and sperm; first cell of a new individual.

1148. zygotic meiosis

The occurrence of meiosis during the division of a zygote in sexually reproducing organisms.

Classical Mendelian Genetics

1149. allele

One of several different genes for a trait that can occupy a particular site on a chromosome.

1150. aneuploidy

Having more or less than the normal diploid number of chromosomes.

1151. autosomal

Concerning paired (nonsex) chromosomes and the genetic information they carry.

1152. autosome

One of a pair of nonsex chromosomes.

1153. codominance

The simultaneous expression of two alleles in one individual without blending.

1154. crossing-over

The exchange of corresponding segments of DNA during meiosis.

1155. deletion

The loss of one or more bases from a DNA strand.

1156. differential migration

Alteration of gene frequencies in a population by population mobility.

1157. dihybrid

Involving two pairs of alleles.

1158. dominant allele

The allele expressed when it and a recessive allele are present.

1159. epistatic

One gene interfering with the effect of another gene.

1160. genetics

The study of heredity.

1161. genotype

Alleles of a single gene or all the genes carried by a particular individual.

1162. hemizygous

Having one of each kind of sex chromosome.

1163. heredity

Transmission of characteristics from one generation to the next.

1164. heterozygous

Having unlike alleles for a trait.

1165. homologous

Having the same shape and structure; pairs of chromosomes having the same information.

1166. homozygous

Having like alleles for a trait.

1167. hybrid

An offspring produced by crossing populations differing in one or more traits.

1168. hybrid vigor

The display of increased fitness resulting from crossing of populations.

1169. incomplete dominance

The expression by blending of two alleles present at the same time in an organism.

1170. linkage

Degree to which genes are closely associated physically and thereby inherited together.

1171. linkage group

A group of genes physically close and inherited together.

1172. monohybrid

Pertaining to a genetic cross that differs in a single trait under study.

1173. mosaic

Organism with some cells containing different genetic information than other cells.

1174. multiple alleles

Three or more kinds of genetic information for a given trait.

1175. phenotype

The appearance of an individual with respect to one or all inherited characteristics.

1176. pleiotropy

The influence of a single gene on more than one trait.

1177. polygenic inheritance

A situation in which two or more genes, each with alleles, jointly affect the expression of a trait.

1178. polyploidy

Presence of more than two sets of chromosomes.

1179. principle of dominance

A Mendelian principle that one factor for a trait can mask or overpower another factor for the same trait.

1180. principle of unit characters

A Mendelian principle that individuals carry two factors for each trait.

1181. Punnett square

A table to illustrate the distribution of alleles in the offspring of heterozygtes.

1182. recessive

In genetics, a characteristic seen in the phenotype only when recessive allele is only one present in the genotype.

1183. selection

Favoring of certain alleles in a gene pool.

1184. sex-linked

Located on a sex chromosome.

1185. testcross

The mating of an organism with an unknown genotype to one with a homozygous recessive phenotype.

1186. tetraploidy

Having four sets of chromosomes.

1187. triploidy

Having three sets of chromosomes.

1188. variation

Divergence among individuals in a species.

Molecular and Population Genetics

1189. anticodon

A three-base sequence of transfer RNA that fits with a particular codon on messenger RNA.

1190. auxotrophic

A mutant organism that has greater nutrional requirements than its normal counterpart.

1191. bacteriophage

A virus that infects bacteria.

1192. biotechnology

The use of a natural biological system to make a product or achieve a particular end.

1193. codon

A three-base sequence in messenger RNA derived from DNA and specifying amino acid placement in a protein.

1194. complementary base pairing

Bonding between certain bases in nucleic acid strands.

1195. DNA ligase

An enzyme that attaches cut ends of DNA molecules.

1196. frameshift mutation

A DNA sequence change caused by adding or deleting bases.

1197. gene

Functional unit of heredity; a site on a chromosome that transmits a particular hereditary characteristic.

1198. gene amplification

Selective replication of certain genes.

1199. gene sequencing

Determination of the order of nucleotides in a gene.

1200. gene therapy

Biotechnical applications designed to treat genetic disease.

1201. genetic engineering

Use of human-designed procedures to alter genetic information.

1202. genetic equilibrium

A state of constancy in the frequency of alleles in a population.

1203. genetic load

The number of genetic defects present in a population.

1204. genetic screening

Search for genetic defects in fetuses, newborns, and prospective parents.

1205. genome

An organism's whole complement of DNA.

1206. genomic library

A collection of genetically engineered viruses carrying all the genes of a species.

1207. inducer

A regulatory molecule that promotes expression of a gene.

1208. inducible enzyme

An enzyme synthesized only in the presence of its substrate.

1209. insertion mutation

A change in DNA involving the addition of nucleotides at some point along the strand.

1210. inversion

A change in a chromosome resulting in a reordering of its genes.

1211. lagging strand

The DNA strand on which synthesis is discontinuous during replication.

1212. leading strand

The DNA strand that is replicated continuously.

1213. missense mutation

A point mutation that replaces one amino acid for another in a protein.

1214. molecular hybridization

A procedure for determining similarity of nucleotide sequence between two nucleic acid molecules.

1215. mutagen

An agent that can alter DNA.

1216. mutation

A change in genetic information.

1217. nonsense mutation

A point mutation that produces a codon that stops protein synthesis, thereby creating a short nonfunctional peptide.

1218. Okazaki fragments

Short segments of single-stranded DNA synthesized on a lagging strand.

1219. oncogene

A gene that contributes to the development of cancer.

1220. operon

A group of bacterial genes that function together and are controlled by a regulator gene.

1221. plasmid

Extrachromosomal DNA that replicates independently in a host cell.

1222. point mutation

A change in a single base in a DNA molecule.

1223. protoplast

A plant cell with its cell wall removed.

1224. purine

A nitrogenous base with two rings found in nucleic acids.

1225. pyrimidine

A nitrogenous base with one ring found in nucleic acids.

1226. recombinant DNA

DNA segments combined from two different organisms.

1227. regulator gene

A gene that directs the activity of a specific set of genes.

1228. replication

Duplication.

1229. repressor

A gene that prevents the action of a set of genes.

1230. restriction enzyme

An endonuclease that cuts double stranded DNA at sites having specific nucleotide sequences.

1231. restriction site

The site at which an endonuclease acts.

1232. reverse transcriptase

An enzyme that makes DNA according to an RNA template.

1233. ribosomal RNA (rRNA)

A nucleic acid that forms part of a ribosome.

1234. semiconservative replication

The replication of DNA in which each molecule consists of one new and one old strand.

1235. structural gene

A gene that produces a specific product.

1236. template

Pattern.

1237. transcription

The transfer of coded genetic information from DNA to mRNA.

1238. transfer RNA

RNA that carries amino acids and places them in specific sites in a growing peptide chain.

1239. translation

The process by which mRNA codons are used to determine the sequence of amino acids in a protein.

1240. translocation

Transfer of part of a chromosome from its normal
location to a location on another chromosome.

1241. vector

A DNA carrier that can insert foreign genetic material into
a host cell.

Development

1242. abscisic acid

A plant hormone that inhibits growth and induces dormancy in buds and seeds.

1243. abscission

Separation and dropping of leaves, fruits, and other plant parts.

1244. acinus

A small cluster.

1245. adipose

Pertaining to fat.

1246. adolescence

Period from the onset of puberty and adulthood.

1247. aging

Process of growing old.

1248. alecithal

Pertaining to an egg having no yolk.

1249. allantois

A fetal membrane that helps to form the umbilical cord.

1250. amniocentesis

Procedure for taking a sample of amniotic fluid from around a fetus to detect genetic or developmental defects.

1251. amnion

Membrane around a fetus that fills with fluid and acts as a shock absorber.

1252. amniote

Organism whose embryos have an amnion and other embryonic membranes.

1253. amniotic

Of the amnion.

1254. anamniote

Animal whose embryos lack embryonic membranes.

1255. apical meristem

Dividing cells found in a terminal bud.

1256. archenteron

A primitive gut cavity in an animal embryo.

1257. auxin

A plant hormone that stimulates cell division and cell growth.

1258. blastocoele

Cavity in a blastula.

1259. blastocyst

Hollow ball of cells that arises early in embryonic development.

1260. blastomere

A cell from early divisions during cleavage of a zygote.

1261. blastula

A hollow ball of cells in the early embryonic development of an animal.

1262. cambium

A sheet of unspecialized cells between xylem and phloem that can divide to produce either tissue.

1263. cell plate

A structure between two plant cells at the end of mitosis where cell membrane and cell wall will develop.

1264. chorion

Outermost fetal membrane, which is incorporated into the placenta.

1265. chorionic villi

Tufts of fetal blood vessels across which substances are exchanged with maternal blood.

1266. chromosomal abnormality

Detrimental change in the DNA configuration in a chromosome.

1267. cleavage

Division into two equal parts; process by which a zygote develops into a multicellular ball.

1268. collagen

A fibrous protein in connective tissue.

1269. collenchyma

Supporting cells with reinforced walls found in elongating stems and petioles of plants.

1270. companion cell

A specialized cell found adjacent to a sieve-tube cell.

1271. congenital

Present at birth.

1272. connective tissue

Tissue of fibrocytes and other cells imbedded in ground substance deposited in an organic fibrous matrix.

1273. differentiation

The specialization of structures during embyronic development.

1274. dilation

Increasing in diameter.

1275. dizygotic

Arising from two separate zygotes.

1276. duct

A tube that usually carries a secretion.

1277. ectoderm

The outermost germ layer in an embryo.

1278. embryonic disc

Trophoblast cells that give rise to the embryo.

1279. embryonic induction

Interaction of two embryonic tissues in which one affects the developmental potential of the other.

1280. endoderm

The innermost germ layer in an embryo.

1281. epigenesis

A theory that the development of an embryo consists of gradual elaboration and organization of parts.

1282. epithelium

A thin tissue that lines hollow organs or covers surfaces.

1283. ethylene

A simple organic molecule that stimulates fruit ripening.

1284. exocrine

Of a gland with ducts.

1285. extraembryonic membrane

One of several membranes that surround a developing vertebrate embryo.

1286. fertilization

The union of egg and sperm.

1287. fertilization membrane

The vitelline membrane after fertilization.

1288. fibroblast

A connective tissue cell that makes fibers and ground substance.

1289. gastrula

The stage of a developing vertebrate embyro that follows the blastula.

1290. gastrulation

The developmental process of forming a gastrula.

1291. gibberellin

A plant hormone that stimulates stem growth and affects flowering, root formation, and leaf growth in some plants.

1292. histology

The study of tissues.

1293. holoblastic cleavage

Undergoing complete cleavage of the zygote.

1294. Huntington's chorea

A dominant hereditary disorder that causes degeneration of the nervous system.

1295. infancy

The period of time from 1 month to 2 years of age.

1296. inner cell mass

Cells that give rise to a mammalian embryo.

1297. intercalated

Inserted between other structures.

1298. internode

The space between nodes on a plant stem.

1299. Klinefelter's syndrome

A condition due to the presence of XXY sex chromosomes.

1300. lacuna

Cavity.

1301. leaf axil

The point at which a leaf and stem unite.

1302. matrix

A fibrous framework in which ground substance of connective tissue is deposited.

1303. meroblastic cleavage

Cleavage restricted to the cytoplasmic part of a yolk laden egg.

1304. mesoderm

The middle germ layer in an embryo.

1305. monozygotic

Arising from the same zygote.

1306. morula

A solid ball of cells in early embryological development.

1307. muscle

A tissue capable of contracting.

1308. neonate

A newborn infant.

1309. nervous tissue

A tissue capable of conducting signals or impulses.

1310. node

A portion of a stem that produces leaves or branches in stems.

1311. nondisjunction

Failure of replicated chromosomes to separate.

1312. notochord

A rigid rod of cells beneath the nerve cord that gives rise to vertebrae.

1313. oligolecital

Pertaining to an egg with a small amount of yolk.

1314. organogenesis

The embryonic development of organs.

1315. palisade cell

Part of a compact layer of cells in the mesophyll of fern and seed plant leaves.

1316. parenchyma

Cells that form packing material inside leaves, stems, and roots.

1317. periderm

Protective tissue that replaces the epidermis in widening stems; cork, cork cambium, and cork parenchyma.

1318. phototrophism

The bending of a plant toward light.

1319. pith

Tissue at the center of the stem or root of a plant.

1320. plumule

The first shoot and leaves of a plant.

1321. pollination

Detrimental alteration of the environment, usually by human activities.

1322. primary tissue

Tissue composed of cells produced by meristem at the tips of roots and stems.

1323. primitive streak

An area in an early embryo having a large amount of yolk that represents the blastopore in embryos with less yolk.

1324. reproductive engineering

Human-designed procedures used to alter the reproductive process.

1325. scab

A crust over a superficial wound.

1326. scar

Connective tissue that has replaced an injured tissue unable to replace itself.

1327. sclerenchyma

A supproting tissue in thick-walled plant cells.

1328. secondary tissue

Tissue originating from the vascular cambium or cork cambium and that increases root or stem diameter.

1329. seed

An embryonic plant consisting of the embryo, stored food, and a coat; a mature ovule.

1330. somite

An embryonic segment of mesoderm.

1331. spore

A reproductive cell capable of producing a new individual without uniting with another cell.

1332. stipule

Paired outgrowths of a stem at the base of a leaf.

1333. stratum

A layer, usually of tissue.

1334. telolecithal

Having an abundant supply of yolk.

1335. trisomy

Condition of having three copies of a chromosome.

1336. trophoblast

Outer blastocyst layer that is connected with maternal tissue and gives rise to chorionic villi.

1337. Turner's syndrome

A condition due to having a single X chromosome (without another X or Y chromosome).

1338. vessel element

A cells comprising a vessel in xylem.

1339. vitelline membrane

A protective covering membrane of an egg.

1340. Wharton's jelly

Soft, pulpy connective tissue of the umbilical cord matrix.

1341. yolk

Food storage part of an egg.

1342. yolk sac

Embryonic bag containing nutrients.

Evolution

1343. abiogenesis

Beginning without life; spontaneous generation.

1344. adaptive radiation

Change in a population over time such as by divergence or convergence.

1345. allopatric speciation

The rise of new species from populations of a single species that were physically isolated from one another.

1346. analogous structure

One of two or more structures having similar functions.

1347. anthropoid

An advanced primate, such as a monkey, ape, or human.

1348. arboreal

Tree-living.

1349. arthropod

An animal with a segmented body and jointed appendages.

1350. binomial system

The two-name system of naming organisms.

1351. biogenesis

Generating life from life.

1352. biological evolution

Changes over time in living organisms.

1353. chemical evolution

The gradual increase in complexity of molecules thought to have preceded the origin of living cells.

1354. coacervate droplet

A mixture of large molecules thought to have preceded the organization of the first cells.

1355. common ancestor

An ancestor to two or more branches in the evolutionary tree.

1356. comparative anatomy

The study of similarities and differences in structure among organisms.

1357. dichotomous key

A means of identifying organisms by chosing which of paired statements in a series pertain to an organism.

1358. directional selection

Selection involving changes that occur when a population displays a steady trend over time.

1359. disruptive selection

Selection due to unusual features that have high survival value.

1360. divergence

Radiating out in different directions.

1361. ecological equivalent

An unrelated organism having functions similar to another in the same environment.

1362. energy

The ability to do work.

1363. evolution

The process of change over time.

1364. evolutionary tree

A diagram showing evolutionary relationships among selected organisms.

1365. extinct

No longer having living representatives.

1366. five kingdom system

A taxonomic system that places all living organisms in one of five kingdoms.

1367. fossil

Any evidence of organisms that lived in the past.

1368. founder effect

Evolutionary effect of a small, nonrepresentative population giving rise to a species; extreme case of genetic drift.

1369. gene flow

The movement of genes from one population to another by reproduction between members of the populations.

1370. gene pool

The sum of all genes and their alleles present in a population at a given time.

1371. genetic drift

Fluctuations in gene frequencies due to isolation of nonrepresentative sampling of the founding population.

1372. genus

The taxonomic category that combines similar species; the first part of the scientific name of an organism.

1373. Hardy-Weinberg equilibrium

A state in which gene frequencies remain unchanged from generation to generation in a population.

1374. hominid

A modern human or an ancestor.

1375. isolating mechanism

A factor that prevents matings between members of two populations.

1376. macroevolution

Large scale evolutionary change.

1377. microsphere

A structure consisting of protein but having certain attributes of a cell.

1378. oxidizing atmosphere

An atmosphere containing oxygen and other oxidizing molecules.

1379. phylogenetic tree

A branching diagram showing evolutionary relationships among selected organisms.

1380. polymorphic

Pertaining to a species having two or more distinct kinds of individuals.

1381. population

A set of all interacting, interbreeding individuals of one species.

1382. primate

A mammal with grasping hands, well developed collar bones, and eyes directed forward.

1383. primitive atmosphere

Atmosphere having gases that were present prior to the emergence of life on earth.

1384. race

An interbreeding population with particular gene frequencies different from those of other such groups.

1385. scientific name

The genus and species to which an organism belongs.

1386. speciation

The creation of a new species.

1387. species

A group of similar organisms having common genes.

1388. sympatric speciation

Origin of a new species from a group of an existing species because of physiological or behavioral isolation.

1389. taxonomy

The science of classifying organisms.

1390. trochophore

A free swimming, ciliated larva.

1391. vestigial structure

The remains of a structure that was once functional in an ancestor.

Viruses and Prokaryotes

1392. ammonification

Conversion of protein to ammonia by soil bacteria.

1393. antibiotic

An organic compound produced by a living organism and used to treat an infection.

1394. archaebacterium

A primitive kind of bacterium.

1395. bacillus

A rod-shaped bacterium.

1396. bactericidal agent

An agent that kills bacteria.

1397. bacteriostatic agent

An agent that inhibits bacterial growth.

1398. binary fission

Asexual reproduction in which a cell or an organism separates into two cells.

1399. capsid

The outer coat of a virus composed of subunits of protein.

1400. chemotherapy

The treatment of cancer or an infectious disease with chemical agents.

1401. coccus

A spherical bacteriuim.

1402. colony

A organism consisting of a loose collection of cells with a modest degree of specialization.

1403. conjugation

A sexual union in which nuclear material of one cell enters another.

1404. diplococcus

A bacterium found in pairs of spherical cells.

1405. elicitor

A substance produced by a plant pathogen that induces the host plant to make resistant phytoalexins.

1406. endospore

A spore formed by certain bacteria that allow them to survive unfavorable conditions.

1407. parasite

An organism that resides in or on another organism and does harm to the host organism.

1408. pathogen

An organism that can cause disease in another organism.

1409. phycobilin

A water soluble pigment found in cyanobacteria and red algae and related to bile pigments in animals.

1410. phycobilisome

A structure on a photosynthetic membrane that contains phycobilins.

1411. retrovirus

A virus that contains RNA and uses it to make DNA.

1412. rickettsia

Microorganisms smaller than bacteria that reproduce only in cells.

1413. spirillum

Bent rod or spiral shaped bacterium.

1414. vibrio

A comma-shaped bacterium.

Protists and Fungi

1415. anisogamy

Having gametes of different sizes.

1416. ascospore

A spore from a single-celled fungus capable of surviving long periods of drought and extreme temperature.

1417. ascus

A sac-shaped spore forming cell associated with sexual reproduction in certain fungi.

1418. basidiospore

A spore produced by basidia in gills of mushrooms.

1419. basidium

A club-shaped structure in which spores are produced.

1420. bioluminescence

The giving off of light by living organisms.

1421. bread mold

A fungus commonly found growing on bread.

1422. coenocyte

A multinucleate organism or one made of multinucleated cells.

1423. conidium

A spore produced by a sac or club fungus during asexual reproduction.

1424. cytostome

A permanent site in a ciliate through which food enters.

1425. dikaryotic

Having two nuclei.

1426. fungus

An organism of the kingdom containing molds and plantlike organisms lacking chlorophyll and feeding on detritus.

1427. gametangium

A gamete producing structure in a plant.

1428. haustorium

A fungal hypha or other plant part that takes up nutrients.

1429. hypha

A threadlike extension of a fungus.

1430. karyogamy

Conjugation of cells and union of nuclei.

1431. lichen

A symbiotic combination of an alga and a fungus.

1432. mycelium

A mass of hyphae making up the body of a fungus.

1433. oogamy

Having gametes differentated as egg and sperm.

1434. oogonium

A mitotically dividing female germ cell that produces primary oocytes.

1435. plasmodium

A slimy mass of acellular cytoplasm.

1436. plasmogamy

Cytoplasmic fusion of cells.

1437. protist

A member of the Kingdom Protista.

1438. protozoa

Primitive motile, single celled organisms.

1439. pseudopod

An extension from the main cell mass of an amoeba.

1440. rhizoid

A root-like organ in moss plants.

1441. saprophyte

An organism that feeds by absorbing molecules from decaying organic matter.

1442. sporangium

A spore bearing structure on the underside of a fern leaf.

1443. tropism

A fixed kind of movement in response to a stimulus.

1444. zooplankton

Animal-like protists usually floating in a body of water.

Plants and Plant Reproduction

1445. adventitious

A structure that arises from other than its normal site.

1446. agar

A complex carbohydrate derived from red algae and used to solidify bacterial growth media.

1447. algal bloom

A population explosion among algae, usually on a pond surface.

1448. algin

A substance found in the cell wall of kelp and other algae and used in certain foods.

1449. angiosperm

A flowering vascular plant with seeds in carpels that develop into fruits.

1450. anther

The part of a stamen in which pollen is formed.

1451. antheridium

A male structure in some nonseed plants in which swimming sperm are produced.

1452. apical dominance

A terminal bud's inhibition of lateral bud growth.

1453. archegonium

A female structure in certain nonseed plants in which an egg is produced.

1454. asexual reproduction

Reproduction not involving the union of gametes.

1455. biennial

Occurring every two years.

1456. bryophyte

A nonvascular plant such as a liverwort or moss.

1457. bud

A short, immature section of a plant stem.

1458. budding

Asexual reproduction in which a new organism or cell pinches of from the parent.

1459. bulb

A compressed stem bearing fleshy leaves.

1460. capillary action

The ability of a liquid to rise against gravity because of molecular cohesiveness and adherence.

1461. chemotropism

The movement of cells or organisms toward or away from certain chemical substances.

1462. coleoptile

A sheath that covers the first leaves of a grass seedling.

1463. complete flower

A flower with sepals, petals, stamens, and a pistil.

1464. corm

A short bulky stem containing stored food.

1465. corolla

The entire whorl of a flower's petals.

1466. cotyledon

A seed leaf in a plant embryo.

1467. cross-pollination

Pollination involving the transfer of pollen from one plant to another.

1468. cycad

A cone-bearing palmlike plant found in the tropics.

1469. cytokinin

A plant hormone associated with cell division.

1470. diatom

A unicellular alga with cell walls containing silica.

1471. dicotyledon

A plant whose embryo had two seed leaves.

1472. dioecious

Having staminate and pistillate flowers on separte plants.

1473. double fertilization

The two fertilizations that form the zygote and endosperm in flowering plants.

1474. endosperm

A nutrient material in a plant embyro.

1475. endosymbiont

An organism that lives with another in a symbiotic association.

1476. epicotyl

Tiny leaves and a bud found inside a typical dicot seed.

1477. ethnobotany

The study of plants as they relate to human culture.

1478. filament

The portion of a stamen that supports an anther.

1479. fruit

A mature ovary of a flowering plant.

1480. gamete

Haploid cell; an ovum or sperm.

1481. gametogenesis

The process of forming gametes.

1482. gametophyte

The haploid, gamete producing portion of a plant life cycle.

1483. gemma

An asexually produced miniature of an organism, typical of bryophytes.

1484. geotropism

A plant's response to gravity.

1485. grafting

A cloning method by which a dormant stem piece is inserted into another stem with a root system.

1486. gymnosperm

One of a group of vascular plants having naked or unenclosed seeds.

1487. herbaceous stem

A soft green stem with little woody tissue.

1488. hypocotyl

A minature embryo stem found inside a typical dicot seed.

1489. incomplete flower

A flower lacking one or more parts that make up a complete flower.

1490. mariculture

Growing of human food in the ocean.

1491. megaspore

Haploid cells, one of which survives to form a female gametophyte in higher plants.

1492. megaspore mother cell

Diploid cell that gives rise to four haploid megaspores.

1493. meristem

Plant tissue containing cells capable of dividing throughout the life of the plant.

1494. microspore

A spore that develops into a pollen grain in seed plants.

1495. microspore mother cell

A cell in the anther that produces four microspores.

1496. monocotyledon

A kind of angiosperm whose embryo has only one seed leaf.

1497. monoecious

Having separate staminate and pistillate flowers on the same plant.

1498. ovule

The part of a plant ovary that develops into a seed.

1499. peat

Partially decomposed deposits of dead mosses.

1500. perennial

Living for several years.

1501. petiole

A structure that connects a leaf to a stem.

1502. phycocyanin

A blue photosynthetic pigment in red algae.

1503. phycoerythrin

A red photosynthetic pigment in red algae.

1504. phytochrome

A plant pigment that responds to red and far red light and participates in photosynthesis.

1505. pistil

The ovary, style, and stigma found together in the center of a flower.

1506. pollen grain

An immature male gametophyte consisting of a generative nucleus and a tube nucleus.

1507. protonema

A body of branching filaments in a spore.

1508. radicle

An embryonic structure in a seed that gives rise to a root.

1509. rhizome

A horizontal underground stem.

1510. root pressure

Force caused by differences in osmotic pressure between the cells of a root hair and cells in xylem or root pith.

1511. runner

A long narrow stem growing horizontally along the surface of the ground; stolon.

1512. scion

In grafting, the stem ctting that is inserted into the stock.

1513. self-pollination

Pollination of a flower by pollen from anthers of the same flower or others on the same plant.

1514. sepal

The outer protective part of a flower.

1515. sorus

A cluster of sporangia.

1516. sporic meiosis

The occurrence of meiosis during spore formation.

1517. sporogenesis

The production of spores.

1518. sporophyll

A sporangium bearing leaf.

1519. sporophyte

The diploid, spore producing generation in alternation of generations.

1520. stamen

The male part of a flower, including filament and anther.

1521. stigma

A sticky structure in the pistil of a flower that receives pollen grains.

1522. stock

In grafting, the rooted stem that supports the scion.

1523. stolon

A long narrow stem growing along the ground; runner.

1524. style

A part of the pistil in a flower that supports the stigma.

1525. tracheophyte

A plant with vascular tissue.

1526. tuber

A bulky terminal part of an underground stem.

1527. turgor

Firmness in cells due to water held in them by osmotic pressure.

1528. vascular plant

A plant having xylem and phloem tissues.

1529. vegetative propagation

Asexual reproduction in flowering plants as from cuttings.

Animals

1530. ametabolous

Lacking metamorphic changes.

1531. amplexus

Clasping and pressing action of a male frog that releases eggs from the female's abdomen.

1532. antenna

A sensory organ on the head of an arthropod.

1533. antennule

An organ of balance and hearing in some arthropods.

1534. appendage

A movable extension of the body, such as a leg or arm.

1535. bilateral symmetry

Symmetry in which a plane can divide equal left and right halves.

1536. biological control

The use of one organism to control another, especially the reduction in numbers of an undesirable organism.

1537. bivalve

Mollusks with two shells hinged dorsally.

1538. carapace

Upper covering or shell on some arthropods.

1539. cephalothorax

The fused head and thorax of a crustacean.

1540. chitin

A flexible substance found in exoskeletons.

1541. chromatophore

A pigmented cell found beneath the outer skin layer in some fishes.

1542. coelom

Body cavity; space between body wall and internal organs.

1543. complete metamorphosis

Development in insects that proceeds through distinct egg, larva, pupa, and adult stages.

1544. crop

A chamber in birds and earthworms where food is temporarily stored.

1545. cysticercus

An encysted developmental stage of a tapeworm.

1546. deuterostome

Member of a large group of animal phyla having an anus derived from a blastopore.

1547. dipleurula

A larval form unique to deuterostomes.

1548. ectotherm

An animal whose body temperature is determined by the environment.

1549. egg tooth

A single central tooth used by reptiles and birds during hatching from an egg and subsequently lost.

1550. endoskeleton

An inner skeleton.

1551. endotherm

An animal whose body temperature is maintained in a narrow range by internal processes.

1552. exoskeleton

An external skeleton.

1553. filter feeder

An animal that removes particulate food from water.

1554. flame cell

A tubular network that constitutes the excretory system of a planarian.

1555. gastrodermis

A layer of cells that lines the body cavity in cnidarians.

1556. gastrovascular cavity

A central digestive cavity with a single opening found in lower animals.

1557. gizzard

A muscular chamber in birds and earthworms in which food is ground into small pieces by small stones.

1558. green gland

A gland that removes wastes from the blood in crayfish.

1559. haltere

A small second set of wings in certain kinds of flies.

1560. hemimetabolous

Undergoing partial metamorphosis, as occurs in some insects.

1561. hemocoel

Blood filled body cavity in animals with an open circulatory system, such as arthropods.

1562. hemocyanin

An oxygen-carrying copper compound in the blood of mollusks.

1563. hermaphroditic animal

An animal having both male and female sex organs.

1564. holometabolous

Undergoing complete metamorphosis.

1565. incomplete metamorphosis

Gradual development, as in some insects, from egg to nymph to adult.

1566. incurrent pore

A pore through which water enters a sponge.

1567. invertebrate

An animal lacking a backbone.

1568. jointed appendage

A movable, bendable structure extending from an animal's body.

1569. medusa

A free-floating inverted polyp seen in cnidarians.

1570. mesoglea

Nonliving jellylike substance separating ectoderm and endoderm in the sponge body wall.

1571. metamerism

Body segmentation along its primary axis, producing a series of homologous parts.

1572. metamorphosis

Transformation of body structure, especially from larva to adult.

1573. molting

Shedding of an exoskeleton.

1574. nacre

A layer of calcium carbonate that lines some mollusk shells.

1575. nematocyst

A threadlike group of stinging cells used by cnidarian to capture prey.

1576. nymph

An immature insect that differs little from one stage to the next.

1577. oviposter

A pointed organ in a female insect used to make a tunnel in the ground into which eggs are deposited.

1578. parapodium

A footlike fleshy lobe on the segments of marine annelids.

1579. parthenogenesis

The development of unfertilized eggs.

1580. pedipalp

A leglike sensory appendage found in spiders.

1581. pericardial sinus

A space surrounding the heart, especially in crayfish.

1582. pineal eye

A light sensitive structure found in the brain of some extinct snakes.

1583. podium

A footlike structure.

1584. polyp

A cnidarian body attached to a structure in its environment.

1585. proglottid

A section of a tapeworm.

1586. protostome

One of many species of animals in which the mouth develops from the blastopore.

1587. pseudocoelom

A fluid-filled space between the mesoderm and the internal organs of an unsegmented round worm.

1588. pupa

A nonfeeding stage following the final larval stage.

1589. queen

The single mother of a colony of honeybees.

1590. radial symmetry

Symmetry in which all wedge-shaped sections around a vertical line are similar.

1591. scolex

The head of a tapeworm.

1592. septum

A wall or partition.

1593. sessile

Remaining stationary.

1594. seta

A bristle that serves in locomotion of segmented worms.

1595. spherical symmetry

Symmetry in which cone-shaped sections are similar.

1596. spinneret

An abdominal appendage that secretes silk in spiders.

1597. symmetry

Similarity of form or arrangement around a point, line, or plane.

1598. tube foot

A hollow structure on the underside of a starfish used in food-getting and locomotion.

1599. tympanum

Sound receptor found in terrestrial animals.

1600. uropod

Abdominal segments in an arthropod that contribute to forming a paddle.

1601. vertebrate

Having a back bone.

1602. visceral mass

Soft internal organs of a mollusk.

1603. worker

A female honeybee that has no reproductive organs.

Ecology and Environment

1604. acid rain

Rain containing acids from atmospheric pollution.

1605. biodegradable

Capable of being decomposed by living organisms.

1606. biological magnification

The concentration of substances in tissues as they are passed up the food chain.

1607. biomass

The total mass of all organisms living in a particular location.

1608. biome

A major area of the earth having certain life forms maintained by the climate of the region.

1609. biotic potential

The maximum population growth rate under ideal conditions.

1610. carbon cycle

A repetitive sequence of chemical processes in which carbon enters and leaves living organisms.

1611. carrying capacity

The number of individuals of a species a particular environment can support indefinitely.

1612. chapparal

A biome containing broad leaved evergreen shrubs in a dense thicket.

1613. climax community

The final community in ecological succession.

1614. clitellum

A dorsal saddle-like swelling that secretes a cocoon in certain segmented worms.

1615. commensalism

The relationship of two species in which one benefits and the other is neither benefitted nor harmed.

1616. competitive exclusion principle

The principle that if two species continue to compete for the exact same resources, one will become extinct.

1617. consumer

An organism of one population that feeds on organisms of other populations within an ecosystem.

1618. contour cultivation

The planting of crops across a slope to slow water runoff.

1619. coral reef

A structure in tropical waters formed by coral skeletons and proving shelter for many species.

1620. crop rotation

A system of changing the crop grown in a field to control pests and improve soil fertility.

1621. deciduous

The quality of losing leaves in winter or drought.

1622. decomposer

An organism that feeds on the remains of other organisms within an ecosystem.

1623. denitrification

The process of converting nitrate to nitrogen gas; a part of the nitrogen cycle.

1624. density-dependent factor

A population control factor that has a greater effect as the population size increases.

1625. density-independent factor

A population control factor that has the same effect regardless of the size of the population.

1626. deposit feeding

Feeding on materials that collect on the bottom of a body of water.

1627. desalinization

Removal of salt, typically from ocean water.

1628. desert

An area receiving 25 cm or less of rain per year.

1629. desertification

The process by which an area of marginally useful farm land becomes a desert by overgrazing or other abuses.

1630. detritus

Nonliving organic matter.

1631. doubling time

Time required for a population to double in size.

1632. ecological niche

The position of an organism in its environment; its diet, predators, habitat, and effects on its environment.

1633. ecology

The study of how living organisms relate to each other and to their environment.

1634. emergent tree

A tree that stands taller than the others in a forest.

1635. environmental impact statement

A statement of the findings of a detailed study of how an activity might affect the environment.

1636. environmental resistance

The carrying capacity of an environment for a species, stated as environment's resistance to population growth.

1637. epiphyte

A plant that grows on another plant.

1638. estuary

A body water containing a mixture of fresh and salt water.

1639. euphotic zone

The surface layer of a body of water to the depth penetrated by light; region in which photosynthesis occurs.

1640. eutrophication

The process of aging and death of organisms in a pond or lake.

1641. exponential growth

Growth that frequently doubles the population size.

1642. food chain

The flow of energy and matter from the environment through organisms within an environment.

1643. food pyramid

A way of depicting the dependence of animals on other organisms in an ecosystem.

1644. food web

A pattern of interconnected food chains.

1645. fossil fuel

Combustible material, such as coal, oil, gas, derived from previously living material.

1646. geometric population growth

Increase in population by a constant percentage in each generation.

1647. grassland

A biome occupied mainly by grasses and having a dry climate.

1648. greenhouse effect

An increase in environmental temperature as carbon dioxide absorbs heat radiated from the earth.

1649. groundwater

Water in soil and in aquifers beneath the soil.

1650. habitat

The place where an organism normally lives.

1651. hazardous waste

Environmental pollutants that endanger living things.

1652. home range

The region to which an animal or a small group of animals normally confines its activities.

1653. humus

Partially decomposed remains of organisms and their wastes.

1654. hydrothermal vent

A volcanic opening on the ocean floor from separation of tectonic plates that spews forth hot water and minerals.

1655. irruptive growth

Population growth characterized by exponential growth, followed by catastrophic population reductions.

1656. lentic

Pertaining to a lake or pond.

1657. lotic

Pertaining to a creek or river (running water) environment.

1658. marine

Pertaining to the ocean.

1659. monoculture

The growing of only one species of crop over a large area.

1660. mortality

Rate at which members of a population die.

1661. mutualism

A relationship between two organisms of different species that benefits both organisms.

1662. natality

The rate at which new individuals are produced in a population.

1663. net productivity

Amount of energy stored per square meter of land surface.

1664. niche

The role an organism plays in its ecosystem.

1665. nitrification

Formation of nitrates by bacteria.

1666. nitrogen cycle

A sequence of repeated reactions that move nitrogen between organisms and their environment.

1667. nitrogen fixation

A process that chemically combines atmospheric nitrogen with other elements.

1668. nuclear waste

Leftover radioactive material from nuclear power plants and other processes involving radiation.

1669. oligotrophic

Pertaining to an environment having little nutrient available to organisms.

1670. ozone shield

A layer of ozone in the upper atmosphere that protects the earth from damaging ultraviolet radiation.

1671. parasitism

A symbotic relationship in which one organism lives at the expense of and does some damage to the host organism.

1672. permafrost

Perennial frozen subsoil in Artic or subarctic regions.

1673. phytoplankton

Small water-dwelling plantlike protists.

1674. pioneer species

The first organisms to become established in an area at the beginning of ecological succession.

1675. plankton

A collection of free-floating, aquatic organisms carried by fresh water or ocean currents.

1676. pollution

The presence of a substances that damages the environment.

1677. population density

The number of one kind of organism in a defined area.

1678. prairie

A temperate grassland biome having rainfall between 25 and 40 cm per year.

1679. predation

The eating of one organism by another.

1680. prey

An organism eaten by a predator.

1681. primary consumer

An organism that consumes plant material.

1682. primary treatment

The first treatment given to sewage.

1683. producer

An organism that can make organic nutrients from inorganic substances in the environment.

1684. replacement reproduction

A reproductive rate that replaces individuals that die and thus maintains a constant population size.

1685. salinization

Deposition of salt.

1686. salt marsh

Coastal grassland that undergoes seasonal flooding.

1687. savanna

A grassland biome that has occasional trees especially in Africa.

1688. secondary consumer

An animal that eats animals that have fed on plants.

1689. secondary treatment

Bacterial decomposition in a sewage treatment plant.

1690. seral stage

Any of the stages of ecological succession in an ecosystem.

1691. sere

Any stage in ecological succession of an ecosystem.

1692. siltation

Deposition of silt.

1693. slash-and-burn agriculture

The practice of cutting and burning trees to prepare land for agriculture.

1694. stratification

The layering of subcommunities as in a soil layers.

1695. strip-cropping

The planting of alternating strips of different crops in which one crop protects the other.

1696. succession

A series of ecological stages by which the community in a particular area gradually changes.

1697. symbiosis

A relationship between interacting two species of organisms.

1698. taiga

A northern forest containing mainly conifer trees.

1699. temperate deciduous forest

A biome with a moderate climate and many large trees that lose their leaves.

1700. temperature inversion

An event that occurs when air near the ground is cooler than air above it.

1701. terracing

Farming on banks of soil built across a slope to reduce erosion.

1702. tertiary treatment

A process that removes toxic substances and excess minerals from sewage effluent.

1703. thermal pollution

Abnormally high temperature produced in an environment.

1704. thermocline

A sharp temperature difference between layers in a body of water that prevents distribution of oxygen and nutrients.

1705. trophic level

A categorization of species in a food web according to how they obtain nutrients.

1706. tropical rain forest

A hot moist environment with a very large number of species.

1707. tundra

A cold biome with permafrost and lacking trees.

1708. water cycle

A sequence of repetitive reactions in which water moves to and from living things.

1709. water table

A level below which aquifers are filled with water.

1710. wetland

Swamp or march that has standing water most of the year.

1711. zonation

Divisions of a natural community according to variations in physical conditions.

Index

abdomen	1	allantois	106
abdominal	1	allele	97
abduction	73	allergen	49
abiogenesis	114	allergy	49
abscisic acid	106	allopatric speciation	114
abscission	106	alternation of generations	89
absorption	28	alveolus	24
absorptive	28	Alzheimer's disease	57
accommodation	57	amenorrhea	89
acetylcholine	57	ametabolous	130
Achilles tendon	73	amino acid	10
acid	6	ammonification	118
acid rain	136	ammonotelic excretion	54
acid-base balance	54	amniocentesis	106
acinus	106	amnion	106
acquired immune deficiency syndrome (AIDS)	49	amniote	106
acquired immunity	49	amniotic	106
acrosome	89	amplexus	130
actin	73	amplitude	57
action	73	ampulla	57
action potential	57	amylase	20
active immunity	49	anabolic	20
active transport	13	anabolic steroid	82
Adam's apple	24	anabolism	20
adaptation	57	anabolism	22
adaptive radiation	114	anaerobic	20
adenohypophysis	57	analogous structure	114
adenosine triphosphate	10	anamniote	107
adhesion	38	anaphase	13
adipose	106	anaphylaxis	49
adolescence	106	anatomy	1
adrenal	82	androgen	82
adrenalin	82	anemia	38
adrenergic	57	anesthetic	57
adrenocorticotropic hormone	82	aneuploidy	97
adsorptive endocytosis	13	aneurysm	38
adventitia	28	angiosperm	123
adventitious	123	anion	6
aerobic	20	anisogamy	120
afferent	57	anorexia nervosa	28
agar	123	antagonist	73
agglutinin	49	antenna	130
agglutinogen	49	antennule	130
aggression	86	anterior	1
aging	106	anther	123
agonist	73	antheridium	123
agranular leukocyte	38	anthropoid	114
albumin	38	antibiotic	118
alcoholism	28	antibody	49
aldosterone	82	anticoagulant	38
alecithal	106	anticodon	101
algal bloom	123	antidiuretic hormone (ADH)	82
algin	123	antigen	49
alkaline	6	anus	28

aorta	38	basal metabolic rate (BMR)	28
apical	24	basal metabolism	28
apical dominance	123	base	6
apical meristem	107	basidiospore	120
aplastic	38	basidium	120
appendage	130	basilar membrane	58
appendicitis	28	basophil	39
aqueous	57	bicuspid	28
arachnoid	58	biennial	123
arbor vitae	58	bilateral	58
arboreal	114	bilateral symmetry	130
archaebacterium	118	bile	29
archegonium	123	binary fission	118
archenteron	107	binding site	13
areola	89	binocular vision	58
argentaffin	28	binomial system	114
arteriole	38	biodegradable	136
artery	38	biofeedback	58
arthropod	114	biogenesis	114
articulation	73	biological clock	86
artificial insemination	89	biological control	130
artificial pacemaker	38	biological evolution	114
artificially acquired active immunity	49	biological magnification	136
artificially acquired passive immunity	49	biology	1
ascorbic acid	28	bioluminescence	120
ascospore	120	biomass	136
ascus	120	biome	136
asexual reproduction	89	biosphere	1
asexual reproduction	123	biotechnology	101
association neuron	58	biotic potential	136
aster	13	bivalve	130
asthma	24	blastocoele	107
astigmatism	58	blastocyst	107
astrocyte	58	blastomere	107
atom	6	blastula	107
atomic number	6	blind spot	59
atomic weight	6	blood	39
atrium	38	Bohr effect	39
atrophy	73	bolus	29
auditory	58	book gill	24
auricle	38	book lung	24
autonomic nervous system	58	botany	1
autosomal	97	Bowman's capsule	54
autosome	97	Boyle's law	24
autotroph	28	brachial	59
auxin	107	brain stem	59
auxotrophic	101	brain wave	59
axon	58	bread mold	120
axon terminal	58	Broca's motor speech area	59
B lymphocyte	50	bronchiole	24
bacillus	118	bronchitis	24
bacterial antagonism	50	bronchus	24
bactericidal agent	118	browser	86
bacteriophage	101	bryophyte	123
bacteriostatic agent	118	bud	123
ball-and-socket joint	73	budding	89
Bartholin's gland	89	budding	124

buffer	6	chemoreceptor	60
bulb	124	chemosynthesis	22
bulimia	29	chemotaxis	50
bursa	73	chemotherapy	118
bursa of Fabricius	50	chemotroph	29
calcification	73	chemotropism	124
calcitonin	82	chiasma	60
callus	73	chitin	130
calorie	29	chlorophyll	22
calyx	54	chloroplast	13
cambium	107	cholinergic	60
capillary	39	cholinesterase	60
capillary action	124	cholinesterase inhibitor	60
capsid	118	chorion	107
carapace	130	chorionic villi	107
carbaminohemoglobin	39	choroid	60
carbohydrate	10	chromatin	13
carbon cycle	136	chromatophore	130
cardiac	39	chromosomal abnormality	107
cardiopulmonary resuscitation (CPR)	24	chromosome	14
cardiovascular	39	chylomicron	29
carnivore	29	chyme	30
carotene	29	chymotrypsin	30
carpal	73	ciliary body	60
carrying capacity	136	cilium	14
cartilage	74	circadian rhythm	86
Casparian strip	39	circle of Willis	60
catabolism	20	circulation	39
catabolism	22	clavicle	74
catalyst	6	cleaning symbiosis	86
cataract	59	clearance	54
catecholamine	59	cleavage	108
cation	6	climax community	136
CCK-PZ (cholecystokinin-pancreozymin)	29	clitellum	136
cecum	29	clitoris	89
celiac	39	clonal selection theory	50
cell	13	clone	50
cell cycle	13	closed circulatory system	39
cell membrane	13	coacervate droplet	114
cell plate	107	coccus	118
cell theory	13	coccyx	74
cell-mediated immunity	50	cochlea	60
cellulose	29	codominance	97
cementum	29	codon	101
central nervous system (CNS)	59	coelom	130
centriole	13	coenocyte	120
cephalization	59	coenzyme	14
cephalothorax	130	cohesion	39
cerebellum	59	coleoptile	124
cerebrospinal fluid	59	collagen	108
cerebrovascular	59	collecting duct	54
cerebrum	59	collenchyma	108
cerumen	60	colloid	6
cervix	89	colloidal dispersion	6
chapparal	136	colon	30
chemical evolution	114	colony	118
chemolithotroph	29	colostrum	89

commensalism	136	crossing-over	97
common ancestor	115	cross-matching	40
community	1	cross-pollination	124
companion cell	108	cupula	61
comparative anatomy	115	cuspid	30
competition	86	cuticle	50
competitive exclusion principle	137	cycad	124
complement	50	cyclic photophosphorylation	22
complementarity	1	cystic fibrosis	30
complementary base pairing	101	cysticercus	131
complete flower	124	cytokinesis	14
complete metamorphosis	131	cytokinin	124
compound	7	cytoplasm	14
concha	74	cytoplasmic streaming	14
conditioning	86	cytoskeleton	14
conduction deafness	60	cytosol	14
conduction system	39	cytostome	120
cone	60	Dalton's law	24
congenital	108	dark reaction	22
conidium	120	data	1
conjugation	118	deamination	20
conjunctiva	60	decibel	61
connective tissue	108	deciduous	137
consciousness	61	deciduous teeth	30
consumer	137	decomposer	137
continuity of life	1	deductive reasoning	1
contour cultivation	137	defecation	30
contractile protein	74	deglutition	30
contractility	74	dehydration	7
contraction cycle	74	deletion	97
conus arteriosus	40	denaturation	7
convergence	61	dendrite	61
coral reef	137	denitrification	137
core body temperature	30	density-dependent factor	137
cork cambium	40	density-independent factor	137
corm	124	dentin	30
cornea	61	deoxyribonuclease	10
corolla	124	deoxyribonucleic acid (DNA)	10
corpus albicans	89	deposit feeding	137
corpus allata	82	dermis	50
corpus callosum	61	desalinization	137
corpus cardica	82	desert	137
corpus luteum	90	desertification	138
cortex	61	detritus	138
cortisol	82	deuterostome	131
cotyledon	124	development	2
countercurrent mechanism	54	diabetes mellitus	30
courtship ritual	86	diapedesis	40
covalent bond	7	diaphragm	74
coxal bone	74	diarrhea	30
cranial	1	diatom	124
craniosacral	61	dichotomous key	115
cranium	74	dicotyledon	125
creatine phosphate	74	differential migration	97
crop	131	differentiation	108
crop rotation	137	digestion	30
cross-bridge	74	dihybrid	97

dikaryotic	120	endosperm	125
dilation	108	endospore	119
dioecious	125	endosymbiont	125
dipleurula	131	endotherm	131
diplococcus	118	energy	115
diploid	90	entrainment	86
directional selection	115	entropy	7
disaccharide	10	environment	2
disruptive selection	115	environmental impact statement	138
distal	2	environmental resistance	138
divergence	115	enzyme	10
diversity	2	eosinophil	40
dizygotic	108	epicotyl	125
DNA ligase	101	epidermis	50
DNA polymerase	14	epigenesis	109
DNA replication	14	epiglottis	31
dominance hierarchy	86	epinephrine	82
dominant allele	97	epiphyte	138
dorsal	2	epistatic	97
dorsal root ganglion	40	epithelium	109
double fertilization	125	erection	90
doubling time	138	erythrocyte	40
duct	108	erythropoiesis	40
duodenum	31	erythropoietin	41
dura mater	61	esophagus	31
ecological equivalent	115	essential amino acid	31
ecological niche	138	essential fatty acid	31
ecology	138	estivation	86
ecosystem	2	estrogen	90
ectoderm	108	estuary	138
ectotherm	131	ethnobotany	125
edema	40	ethology	86
efferent	61	ethylene	109
egg tooth	131	eukaryotic	14
ejaculation	90	euphotic zone	138
electrocardiogram (ECG)	40	Eustachian tube	25
electroencephalogram (EEG)	40	eutrophication	138
electrolyte	54	evaporation	31
electron	7	evolution	115
electron transport system	20	evolutionary tree	115
eleidin	50	excitability	2
element	7	excitation-contraction coupling	74
elicitor	119	excretion	54
embryonic disc	108	exergonic	7
embryonic induction	108	exocrine	109
emergent tree	138	exoskeleton	131
emotion	61	expiration	25
emphysema	25	exponential growth	138
emulsification	31	extensibility	75
enamel	31	external	2
endergonic	7	external fertilization	90
endocardium	40	exteroceptor	61
endocrine	82	extinct	115
endoderm	109	extracellular	14
endodermis	40	extraembryonic membrane	109
endoplasmic reticulum	14	facilitated diffusion	15
endoskeleton	131	fallopian tube	90

fascia	75	gastric	31
fatigue	75	gastrin	83
fatty acid	10	gastrodermis	131
feces	31	gastrovascular cavity	132
feedback	2	gastrula	109
female pronucleus	90	gastrulation	109
fenestrated	54	gel	15
fermentation	20	gemma	125
ferritin	41	gene	101
fertilization	109	gene amplification	101
fertilization membrane	109	gene flow	116
fertilized ovum	90	gene pool	116
fetus	90	gene sequencing	101
fever	50	gene therapy	101
fibrin	41	genetic code	15
fibrinogen	41	genetic drift	116
fibroblast	109	genetic engineering	102
fibula	75	genetic equilibrium	102
filament	125	genetic load	102
filter feeder	131	genetic screening	102
filtration	15	genetics	97
fission	90	genitalia	91
five kingdom system	115	genome	102
flagellum	15	genomic library	102
flame cell	131	genotype	98
flavin adenine dinucleotide (FAD)	20	genus	116
flexion	75	geometric population growth	139
fluid regulation	54	geotropism	126
fluid-mosaic model	15	germinal epithelium	41
folacin	31	gibberellin	109
follicle-stimulating hormone (FSH)	83	gill	25
fontanel	75	gingiva	32
food chain	139	gizzard	132
food pyramid	139	glans	91
food web	139	glaucoma	62
fossil	115	globin	41
fossil fuel	139	globulin	41
founder effect	116	glomerulus	54
fovea centralis	62	glottis	25
fracture	75	glucagon	83
frameshift mutation	101	glucocorticoid	83
frequency	62	gluconeogenesis	20
frontal	75	gluten	32
fruit	125	glycine	10
fulcrum	75	glycogenesis	20
functional group	10	glycogenolysis	20
fundus	31	glycolipid	10
fungus	120	glycolysis	21
gallbladder	31	glycoprotein	10
gametangium	120	Golgi apparatus	15
gamete	90	gonad	91
gamete	125	gonorrhea	91
gametogenesis	90	graafian follicle	91
gametogenesis	125	gradient	15
gametophyte	125	grafting	126
ganglion	62	gram molecular weight	7
gas exchange	25	granular leukocyte	41

granum	22	homologous	98
grassland	139	homozygous	98
gray matter	62	hormone	83
grazer	87	humerus	75
green gland	132	humoral immunity	51
greenhouse effect	139	humus	139
groundwater	139	Huntington's chorea	110
growth	2	hybrid	98
guard cell	22	hybrid vigor	98
gustatory	62	hybridoma	51
guttation	41	hydrogen bond	7
gymnosperm	126	hydrolysis	7
gyrus	62	hydrophilic	7
habitat	139	hydrophobic	8
habit-forming	62	hydrostatic pressure	15
habituation	62	hydrothermal vent	139
hair cell	62	hyperosmotic	15
haltere	132	hypersensitivity	51
haploid	15	hypertension	42
Hardy-Weinberg equilibrium	116	hyperthermia	32
haustorium	121	hypertonic	15
hazardous waste	139	hypha	121
heart failure	41	hypocotyl	126
heart murmur	41	hypophysis	62
hematocrit	41	hyposmotic	15
hematopoiesis	41	hypotension	42
heme	42	hypothalamus	62
hemimetabolous	132	hypothermia	32
hemizygous	98	hypothesis	2
hemocoel	132	hypotonic	16
hemocyanin	132	IgA	51
hemodialysis	42	IgD	51
hemoglobin	42	IgE	51
hemolysis	42	IgG	51
hemophilia	42	IgM	51
hemopoiesis	42	ileum	32
hemorrhage	42	ilium	75
Henry's law	25	immune	51
herbaceous stem	126	immunity	51
herbivore	32	immunization	51
heredity	98	immunodeficiency	52
hermaphroditic animal	132	immunoglobulin	52
heterogeneity	51	immunology	52
heterotroph	32	immunosuppression	52
heterozygous	98	immunotoxin	52
hibernation	87	implantation	91
high density lipoprotein (HDL)	32	imprinting	87
hippocampus	62	incisor	32
hirudin	42	inclusive fitness	87
histamine	51	incomplete dominance	98
histology	109	incomplete flower	126
holoblastic cleavage	110	incomplete metamorphosis	132
holometabolous	132	incurrent pore	132
home range	139	incus	75
homeostasis	2	individual fitness	87
homeostatic system	2	individual space	87
hominid	116	inducer	102

151

inducible enzyme	102	Krebs cycle	21
inductive reasoning	3	Kupffer's cell	33
infancy	110	labia	91
inflammation	52	labyrinth	63
ingestion	32	lacrimal	63
innate	63	lactase	33
inner cell mass	110	lactation	91
innervation	63	lacteal	33
insectivorous	32	lactiferous	91
insertion	75	lacuna	110
insertion mutation	102	lagging strand	102
inspiration	25	lanugo	52
insulin	83	larynx	25
intercalated	110	latent learning	87
interferon	52	latent period	76
interleukin	52	lateral	3
internal	3	lateral inhibition	63
internal environment	3	lateral line organ	63
internal fertilization	91	lateralization	63
interneuron	63	law of adequate stimulus	63
internode	110	leading strand	102
interphase	16	leaf axil	110
interstitial	3	leaf vein	43
intracellular	16	learned behavior	87
intrinsic	16	learning	63
intrinsic factor	32	lecithin	10
inversion	102	lens	64
invertebrate	132	lentic	140
ion	8	lenticel	25
ionic bond	8	leukocyte	43
iris	63	lichen	121
irruptive growth	140	ligament	76
ischemia	42	ligand	16
ischium	76	light reaction	22
islet of Langerhans	83	limbic system	64
isogamy	91	linkage	98
isolating mechanism	116	linkage group	98
isomer	8	lipase	33
isometric	76	lipid	11
isosmotic	16	lipoprotein	21
isotonic	16	liter	3
isotope	8	loop of Henle	55
isovolumetric	42	lotic	140
isthmus	63	low density lipoprotein (LDL)	33
jaundice	42	luteinizing hormone (LH)	83
jejunum	32	luteum	91
joint	76	lymph	43
jointed appendage	132	lymph node	43
karyogamy	121	lymphatic	43
karyotype	16	lymphocyte	43
kilocalorie	33	lymphoid	43
kin selection	87	lymphokine	52
kinesthetic	63	lysosome	16
kinetic	8	lysozyme	64
kingdom	3	macroevolution	116
kinin	52	macromolecule	3
Klinefelter's syndrome	110	macronutrient	33

macrophage	52	microfilament	17
macula	64	microglia	64
male pronucleus	91	micronutrient	33
malignancy	16	microscope	3
malleus	76	microsphere	116
malnutrition	33	microspore	126
Malphighian tubule	55	microspore mother cell	126
maltase	33	microtubule	17
mammary gland	92	microvillus	34
mantle	25	mimicry	87
mantle cavity	25	mineral	34
mariculture	126	mineralocorticoid	83
marine	140	missense mutation	102
marrow	76	mitochondrion	17
mass	3	mitosis	17
matrix	110	mixed nerve	65
maxilla	76	mixture	8
mechanoreceptor	64	modiolus	65
median	3	molar	34
medulla	64	mole	8
medulla oblongata	64	molecular hybridization	103
medusa	133	molecule	8
megaspore	126	molting	133
megaspore mother cell	126	monocotyledon	126
meiosis	92	monoculture	140
melanin	52	monocyte	43
membrane potential	64	monoecious	127
memory	64	monohybrid	98
menarche	92	monosaccharide	11
Meniere's disease	64	monosynaptic	65
meninges	64	monozygotic	110
menopause	92	mons pubis	92
menstrual cycle	92	mortality	140
menstruation	92	morula	111
meristem	126	mosaic	98
meroblastic cleavage	110	motility	34
mesencephalon	64	motor end plate	76
mesenchyme	76	motor neuron	65
mesentery	33	motor unit	76
mesoderm	110	mucosa	34
mesoglea	133	multiple alleles	99
mesonephros	55	muscle	111
mesophyll	43	mutagen	103
messenger RNA	16	mutation	103
metabolic rate	33	mutualism	140
metabolism	21	mycelium	121
metabolism	22	myelin	65
metacarpal	76	myocardium	43
metamerism	133	myofibril	77
metamorphosis	133	myofilament	77
metanephridium	55	myoglobin	77
metanephros	55	myoneural junction	77
metaphase	16	myosin	77
metastasis	16	myotome	77
metatarsal	76	nacre	133
meter	3	natality	140
micelle	33	natural selection	3

naturally acquired active immunity	53	oligotrophic	141
naturally acquired passive immunity	53	omnivore	34
negative feedback	83	oncogene	103
nematocyst	133	oncotic pressure	43
neocortex	65	oocyte	92
neonate	111	oogamy	121
nephridium	55	oogenesis	92
nephron	55	oogonia	92
nerve	65	oogonium	121
nerve net	65	open circulatory system	43
nervous tissue	111	operant conditoning	87
net filtration pressure	55	operculum	25
net productivity	140	operon	103
net protein utilization	34	opsin	66
neural crest	65	optic chiasma	66
neural oscillator	65	optic disk	66
neural tube	65	organ	4
neurilemma	65	organ of Corti	44
neurofibrillary tangles	65	organelle	17
neuroglia	66	organic	8
neurohypophysis	83	organism	4
neuromuscular	77	organizational complexity	4
neurotransmitter	66	organogenesis	111
neurulation	66	orgasm	92
neutron	8	origin	77
neutrophil	43	osmolarity	17
niacin	34	osmosis	17
niche	140	osmotic pressure	17
nicotinamide adenine dinucleotide (NAD)	21	ossification	77
Nissl granule	66	osteon	77
nitrification	140	otolith	66
nitrogen cycle	140	oval window	67
nitrogen fixation	140	ovary	92
nociceptor	66	oviduct	92
node	111	oviparous	93
node of Ranvier	66	oviposter	133
noncyclic photophosphorylation	22	ovoviviparous	93
nondisjunction	111	ovulation	93
nonpolar	8	ovule	127
nonsense mutation	103	ovum	93
noradrenalin	66	oxidation	9
norepinephrine	66	oxidative phosphorylation	21
notochord	111	oxidizing atmosphere	116
nuclear	17	oxygen debt	77
nuclear waste	141	oxyhemoglobin	44
nucleic acid	11	oxytocin	67
nucleolus	17	ozone shield	141
nucleoplasm	17	pacemaker	44
nucleotide	11	Pacinian corpuscle	67
nucleus	8	pair bond	87
nutrition	34	palate	26
nymph	133	palisade cell	111
obesity	34	pancreas	34
occipital	77	pantothenic acid	34
Okazaki fragments	103	Pap smear	93
olfactory	66	papilla	53
oligolecital	111	parapodium	133

parasite	119	phylogenetic tree	116
parasitism	141	physiological dependence	67
parasympathetic division	67	physiology	4
parathormone	67	phytochrome	127
parathyroid glands	67	phytoplankton	141
parenchyma	111	pia mater	68
Parkinson's disease	67	pilus	53
parthenogenesis	133	pineal eye	134
partial pressure	26	pineal gland	68
parturition	93	pinna	68
passive immunity	53	pioneer species	141
passive transport	17	pistil	127
patella	77	pitch	68
pathogen	119	pith	112
pattern generator	67	placenta	93
peat	127	plankton	141
pedipalp	133	plantar	78
penis	93	plaque	44
pepsin	34	plasma	44
peptide bond	11	plasma cell	53
perception	67	plasma membrane	18
perennial	127	plasmid	103
pericardial sinus	133	plasmin	44
pericardium	44	plasminogen	44
pericycle	44	plasmodium	121
periderm	111	plasmogamy	121
perikaryon	67	plasmolysis	18
peripheral	4	plasticity	78
peristalsis	35	platelet	44
peritoneum	35	pleiotropy	99
peritubular	55	plexus	68
permafrost	141	plumule	112
pernicious anemia	44	podium	134
peroxisome	17	point mutation	103
petiole	127	polar body	93
Peyer's patch	35	polar compound	9
pH	9	pollen grain	127
phagocytosis	53	pollination	112
phalanges	78	pollution	141
pharynx	26	polygenic inheritance	99
phenotype	99	polymer	11
pheromone	87	polymorphic	116
phloem	44	polyp	134
phospholipid	11	polypeptide	11
phosphorylation	21	polyploidy	99
photoauxotroph	35	polysaccharide	11
photolithotroph	35	pons	68
photon	67	population	117
photoperiodism	88	population density	141
photophosphorylation	22	positive feedback	83
photoreceptor	67	posterior	4
photosynthesis	23	postganglionic	68
phototrophism	112	postovulatory	93
phycobilin	119	postsynaptic	68
phycobilisome	119	potential energy	9
phycocyanin	127	prairie	141
phycoerythrin	127	predation	141

preganglionic	68	radiation	9
preovulatory	93	radicle	127
prepuce	93	radius	78
pressure	26	raphe	78
presynaptic	68	rapid-eye-movement (REM) sleep	69
prey	141	rarefaction	69
primary consumer	142	Rathke's pouch	84
primary follicle	93	reactant	9
primary tissue	112	receptor	18
primary treatment	142	recessive	99
primate	117	recombinant DNA	103
prime mover	78	rectum	35
primitive atmosphere	117	reduction	9
primitive streak	112	reflex	69
primordial	94	refraction	69
principle of dominance	99	regulator gene	104
principle of forward conduction	68	relaxin	84
principle of unit characters	99	releaser	88
producer	142	remission	18
progesterone	83	renin	55
proglottid	134	renin-angiotensin mechanism	55
prokaryotic	18	replacement reproduction	142
prolactin	84	replication	104
proliferative phase	94	repressor	104
pronephros	55	reproduction	4
prophase	18	reproductive efficiency	94
proprioceptor	68	reproductive engineering	112
prostaglandin	84	resistance	45
protein	11	respiration	26
prothrombin	44	respiratory center	26
protist	121	respiratory pigment	26
proton	9	response	69
protonema	127	responsiveness	4
protoplasm	18	restriction enzyme	104
protoplast	103	restriction site	104
protostome	134	reticular activating system (RAS)	69
protozoa	121	retina	69
proximal	4	retinene	69
pseudocoelom	134	retrovirus	119
pseudopod	121	reverse transcriptase	104
puberty	84	rhizoid	121
pubis	78	rhizome	127
pulmonary	26	rhodopsin	69
pulp cavity	35	riboflavin	35
pulse	45	ribonuclease	35
Punnett square	99	ribonucleic acid (RNA)	11
pupa	134	ribosomal RNA (rRNA)	104
purine	103	ribosome	18
Purkinje cell	68	rickets	78
Purkinje fiber	45	rickettsia	119
pus	53	rod	69
pylorus	35	root	69
pyrimidine	103	root cap	45
pyrogen	35	root pressure	128
queen	134	rotation	78
race	117	round window	69
radial symmetry	134	ruminant	35

runner	128	sexual reproduction	94
SA node	45	sickle cell anemia	45
sacrum	78	sieve plate	45
sagittal	4	sieve tube	45
salinization	142	siltation	142
salivary	35	sinus	79
salt marsh	142	slash-and-burn agriculture	142
saltatory	69	sliding filament theory	79
saprophyte	122	smooth muscle	79
sarcomere	78	sociobiology	88
sarcoplasm	78	sodium-potassium pump	18
sarcoplasmic reticulum	78	sol	18
saturated fatty acid	11	solute	18
saturation	11	solution	19
savanna	142	solvent	19
scab	112	soma	70
scapula	79	somatic	70
scar	112	somatostatin	84
Schwann cell	70	somatotropin	84
scientific method	4	somite	112
scientific name	117	sorus	128
scion	128	speciation	117
sclerenchyma	112	species	117
sclerotome	79	specific heat	9
scolex	134	specificity	12
scrotum	94	spectrin	45
scurvy	36	sperm	94
sebum	53	spermatic cord	94
secondary consumer	142	spermatid	94
secondary tissue	112	spermatocyte	95
secondary treatment	142	spermatogenesis	95
secretin	36	spermatogonium	95
secretion	9	spermatozoan	95
seed	112	spermiogenesis	95
selection	99	spherical symmetry	135
selectively permeable	18	sphincter	36
self-pollination	128	sphygmomanometer	45
semen	94	spicule	79
semicircular canal	70	spindle fiber	19
semiconservative replication	104	spinneret	135
seminal vesicle	94	spiracle	26
seminiferous tubule	94	spirillum	119
sensation	70	spirometry	26
sensitization	53	spongy parenchyma	23
sensory	70	sporangium	122
sepal	128	spore	113
septum	134	sporic meiosis	128
seral stage	142	sporogenesis	128
sere	142	sporophyll	128
serotonin	70	sporophyte	128
Sertoli's cell	94	stamen	128
serum	45	stapes	79
sessile	134	static equilibrium	70
seta	134	statocyst	70
sex chromosome	18	stele	45
sex-linked	99	stenosis	45
sexual dimorphism	94	stereoisomer	12

stereotyped behavior	88	taste bud	71
sterility	95	taxis	88
steroid	12	taxonomy	117
stethoscope	46	tear	71
stigma	128	telolecithal	113
stimulus	70	telophase	19
stipule	113	temperate deciduous forest	143
stock	128	temperature inversion	143
stolon	129	template	104
stoma	23	temporal	80
stomodeum	36	tendon	80
stratification	143	tension	80
stratum	113	teratogen	19
streptokinase	46	terracing	143
stress	84	territoriality	88
stressor	84	territory	88
stretch reflex	70	tertiary treatment	143
striation	79	testcross	99
stridulation	88	testis	95
strip-cropping	143	testosterone	84
stroke volume	46	tetanus	80
stroma	95	tetrad	95
structural gene	104	tetraploidy	99
style	129	tetrapod	26
subconscious	70	thalamus	71
suberin	46	theca	95
submucosa	36	theory	4
succession	143	theory of immune surveillance	53
sucrase	36	thermal pollution	143
sulcus	79	thermocline	143
summation	79	thermogenesis	36
surface tension	19	thermoreceptor	71
surface-to-volume ratio	19	thiamine	36
surfactant	26	thirst	55
suture	79	thorax	4
swim bladder	26	thrombin	46
symbiosis	143	thrombocyte	46
symmetry	135	thrombus	46
sympathetic chain ganglion	70	thylakoid	23
sympathetic division	71	thymosin	84
sympatric speciation	117	thymus gland	84
symphysis	79	thyroid gland	71
synapse	71	thyroid-stimulating hormone (TSH)	84
synapsis	95	thyroxine	85
synaptic	71	tibia	80
syncytium	79	tissue	5
synergist	80	tissue plasminogen activator (tPA)	46
syngamy	95	tissue thromboplastin	46
synovial joint	80	tocopherol	36
syphilis	95	tolerance	71
systole	46	tonicity	19
T cell	53	tonsil	27
T lymphocyte	46	tonus	80
T wave	46	torr	46
taiga	143	total parental nutrition (TPN)	36
target cell	71	trace element	9
tarsal	80	trachea	27

158

tracheid	47	valence	9
tracheophyte	129	valve	47
tract	85	variation	100
transamination	21	vas deferens	96
transcription	104	vascular	47
transducin	71	vascular bundle	47
transduction	71	vascular cambium	47
transfer RNA	104	vascular cylinder	47
transferrin	47	vascular plant	129
translation	104	vascular system	47
translocation	105	vector	105
transpiration	47	vegetative propagation	129
transpiration-cohesion theory	47	vein	47
transverse	5	vena cava	48
transverse (T) tubule	80	venous	48
tricuspid	47	ventilation	27
triglyceride	12	ventral	5
triploidy	100	ventricle	48
trisomy	113	venule	48
trochlea	80	vertebra	81
trochophore	117	vertebrate	135
trophic level	143	very low density lipoprotein (VLDL)	36
trophoblast	113	vessel element	113
tropic	85	vestibule	72
tropical rain forest	143	vestigial structure	117
tropism	122	vibrio	119
tropomyosin	80	villus	36
troponin	80	viscera	5
trypsin	36	visceral mass	135
tube foot	135	visceroceptor	72
tuber	129	viscosity	48
tubulin	19	visual area	72
tumor necrosis factor	19	vital capacity	27
tundra	144	vitamin A	37
tunica	47	vitamin D	37
turgor	129	vitamin E	37
turgor pressure	19	vitamin K	37
Turner's syndrome	113	vitelline membrane	113
turnover	21	vitreous	72
twitch	81	viviparous	96
tympanic membrane	71	vocal	27
tympanum	135	vomer	81
ulna	81	Wallerian degeneration	72
umbilicus	96	water balance	56
unsaturated fatty acid	12	water cycle	144
ureotelic excretion	56	water table	144
ureter	56	weight	5
urethra	56	wetland	144
uricotelic excretion	56	Wharton's jelly	113
uridine triphosphate (UTP)	12	white matter	72
urinary bladder	56	withdrawal reflex	72
uropod	135	worker	135
uterine	96	xylem	48
uterine tube	96	yolk	113
uterus	96	yolksac	113
utricle	72	zonation	144
vagina	96		

For more information on other publications by Ed Creager, please visit the author's site...

www.edcreager.blogspot.com

... and please note the author's "signature book" entitled,

**"The Money-Saving Idea Book:
Inside Tips for Starving Students,
Frugal Seniors and Every Financial Survivor"**

www.ingramcontent.com/pod-product-compliance
Lightning Source LLC
Chambersburg PA
CBHW081124170526
45165CB00008B/2541